Enhancing the Productivity of Rainfed Agro-Ecosystem through Suitable Interventions

Enhancing the Productivity of Rainfed Agro-Ecosystem through Suitable Interventions

Edited By

Dr. A.K. Kothari
Chief Scientist and Professor, Soil & Water Engineering
Dryland Farming Research Station, Bhilwara
(Maharana Pratap University of Agriculture & Technology, Udaipur)

Dr. J.K. Balyan
Assistant Professor, Agronomy
Dryland Farming Research Station, Bhilwara
(Maharana Pratap University of Agriculture & Technology, Udaipur)

Dr. R.K. Sharma
Assistant Professor, Soil Science
Dryland Farming Research Station, Bhilwara
(Maharana Pratap University of Agriculture & Technology, Udaipur)

Dr. Om Prakash
Principal Scientist
Division of Agricultural Extension
ICAR-IARI, Pusa, New Delhi

International Books & Periodical Supply Service
(Publisher of Scientific Books)
38, NISHANT KUNJ, PITAM PURA MAIN ROAD
DELHI-110034 (India)
Phone : 011-27352078, Mobile : 09810146811, E-mail : hkjain1975@yahoo.com

Published by :

International Books & Periodical Supply Service

(Publisher of Scientific Books)
38, NISHANT KUNJ, PITAM PURA MAIN ROAD
DELHI-110034 (India)
Phone : 011-27352078, Mobile : 09810146811, E-mail : hkjain1975@yahoo.com

ISBN 978-81-939065-1-4

Composed, Designed & Printed in India

भारतीय कृषि अनुसंधान परिषद
कृषि अनुसंधान भवन-II, पूसा, नई दिल्ली 110 012
INDIAN COUNCIL OF AGRICULTURAL RESEARCH
KRISHI ANUSANDHAN BHAVAN-II, PUSA, NEW DELHI 110 012

डा. नरेन्द्र सिंह राठौड़
उप महानिदेशक (कृषि शिक्षा)
Dr. Narendra Singh Rathore
Deputy Director General (Agril. Edn.)

Phone : 011-25841760 (O)
Fax : 011-25843932
E-mail : ddgedn@gmail.com
nsrdsr@gmail.com
Website : www.icar.org.in

Dated: 10 September, 2018

Foreword

Rainfed agriculture occupy an important role in Indian agriculture as about 68% of country's gross cropped area is under rainfed condition and supports about 40% of human population and 60% of livestock. Rainfed areas contribute about 42% of country's food basket. It is the rainfed belt where cultivation of coarse cereals (91%) oilseeds (80%) and cotton (65%) predominates. Maintaining India's fod security is challenging task since the productivity in irrigated areas is now plateauing, bulk of rising food demand in the country has to be met by enhancing the productivity of rainfed areas. Hence, the crop production needs to be enhanced from the present 0.8-1.0 t/ha to 2.0 t/ha under rainfed area.

Efforts made to enhance the productivity of rainfed farmers through suitable technological interventions on different aspects viz.; resource characterization, rainwater management, crops and cropping systems, nutrient management, energy management, farm mechanization, alternate land use management, climate resilient integrated farming system module, weather forecasting, dynamic crop planning, real time ontingent crop planning and management of common pool resources to improve sustainability of dryland farmers.

I congratulate Dr A.K. Kothari, Professor and Chief Scientist, DFRS, Arjia, Bhilwara and associates have complied this useful publication "Enhancing the Productivity of Rainfed Agro ecosystem through Suitable Interventions". This exellent publication will be useful for researchers, extension functionaries and farmers.

N.S. Rathore

प्रो. उमा शंकर शर्मा
कुलपति
Prof. U.S. Sharma
Vice-Chancellor

Maharana Pratap University of Agriculture and Technology
University Campus, Udaipur-313 001 (Rajasthan), India

महाराणा प्रताप कृषि एवं प्रौद्योगिकी विश्वविद्यालय
विश्वविद्यालय परिसर, उदयपुर–313 001 (राजस्थान), भारत

Ph. (O) : 91-294-2471101 (M) : 09414158224, 09414170239
(F) : 91-294-2470682 E-mail: vc_mpuat@yahoo.co.in; vc@mpuat.ac.in

Foreword

Rainfed agriculture with its complex, diverse and risk prone nature has been a challenge for both researchers and extension functionaries since time immemorial. The central issue here is neither a particular form of farming nor a technology package, but the capacity of the people of these regions to support livelihood of 40 per cent of human beings and 60 per cent of livestock. Massive research and development efforts have been made to achieve both higher productivity and sustainability despite vagaries of monsoon.

All India Co-ordinated Research Project on Dryland Agriculture, Dryland Farming Research Station (DFRS), Arjia, Bhilwara is one of the 23 centers working for generating the improved rainfed technologies viz.; resource characterization, rainwater management, crops and cropping systems, nutrient management, energy management, alternate land use management, climate resilient integrated farming system module and contingent crop planning to enhance the productivity of rainfed agro-ecosystem on sustainable basis since 1984. Therefore, Chapters from 19 eminent scientists from outside the University and 12 eminent scientists from the University have been invited for compilation of this book entitled "Enhancing the Productivity of Rainfed Agro Ecosystem through Suitable Interventions" to summarize different interventions for enhancing sustainable productivity of rainfed area throughout the country.

I congratulate Dr A.K. Kothari, Professor and Chief Scientist, DFRS, Arjia, Bhilwara and editors for bringing out this useful publication, which will help the researchers, extension functionaries and farmers to improve the productivity of rainfed agriculture in India. The principles, philosophies, and technologies presented in this publication have the potential to contribute to improve food security and livelihoods for the people of rainfed regions of the world.

(U.S. Sharma)
Vice-Chancellor

Preface

Rainfed agriculture occupy an important role in Indian agriculture as about 68% of country's gross cropped area is under rainfed condition and supports about 40% of human population and 60% of livestock. Rainfed areas contribute about 42% of country's food basket. It is the rainfed belt where cultivation of coarse cereals (91%), oilseeds (80%) and cotton (65%) predominates. Maintaining India's food security is challenging task since the productivity in irrigated areas is now plateauing, bulk of rising food demand in the country has to be met by enhancing the productivity of rainfed areas. Hence, the crop production needs to be enhanced from the present 0.8-1.0 t/ha to 2.0 t/ha under rainfed area. It has received less attention as compared to the irrigated agriculture due to poor resources and rainfed farming in India is looking for a new horizon because traditional methods of production are inadequate to meet the new challenges. The knowledge so acquire is to be diffused among the researchers, scientists and teachers of ICAR and SAUs to enhance sustainable productivity of agriculture from rainfed areas throughout the country.

Keeping this in view, a 21 days ICAR sponsored Winter School on "Enhancing the Productivity of Rainfed Agro Ecosystem through Suitable Interventions" was organized at the Dryland Farming Research Station (Maharana Partap University of Agriculture and Technology, Udaipur), Arjia, Bhilwara, Rajasthan from 14th December, 2017 to 3rd January, 2018. The book deals with different aspects of rainfed agriculture viz.; resource characterization, rainwater management, crops and cropping systems, nutrient management, energy management/farm mechanization, alternate land use management, climate resilient integrated farming system module, watershed management, conservation agriculture, micro-irrigation and fertigation, GIS based landuse planning, ground water recharge, drought management, weather forecasting, dynamic crop planning, real time contingent crop planning, high-tech agriculture, production of medicinal and aromatic plants, ITKs for rainfed areas ,management of common pool resources to improve sustainability of dryland farmers and Govt. policies for improving and promotion of agriculture in rainfed areas.

We profusely thank Dr. N.S. Rathore, DDG (Education), ICAR and Dr. MB Chetti, ADG (HRD), ICAR for their kind encouragement by supporting in organizing this 21 days Winter School. As editors of this book, we would like to thank all the authors for their efforts and co-operation in bringing out this book in time. Also we thank all the participants of this Winter School for sparing time and interest. We thank to all the staff of DFRS, Bhilwara who helped directly or indirectly in organizing this Winter School. We also thank Ms. Devyani Dodia and Mr. Devendra Kumar Mehta for editorial assistance. We firmly believe that this publication will be highly useful, in one way or the other, for researchers, academicians, extension workers, policy, makers, planners, officials in development institutions/agencies, producers and students.

Editors

List of Contributors

Dr. Narendra Singh Rathore
DDG (Education) - ICAR,
New Delhi, India

Dr. V.M. Mayande
Former Vice Chancellor, Dr. PDKV,
Akola, Maharashtra, India

Dr. P.L. Maliwal
Emeritus Professor (ICAR), MPUAT,
Udaipur, Rajasthan, India

Dr. Virendra Kumar Srivastava
Former Professor and Head,
Deptt. of SWE MPUAT, Udaipur,
Rajasthan, India

Dr. R.A. Kaushik
Professor, MPUAT
Udaipur, Rajasthan, India

Dr. R.C. Purohit
Director
JNU, Jaipur, Rajasthan, India

Dr. S.K. Sharma
Director
Centre for Advanced Faculty
Training on Organic Farming,
Udaipur, Rajasthan, India

Dr. Ram Sakal Singh
Regional Director, NBSS & LUP,
Udaipur, Rajasthan, India

Dr. M.S. Shirahatti
Chief Scientist, Regional Agricultural
Research Station, Vijayapura,
Karnataka, India

Dr. V.S. Surakod
AICRP for Dryland Agriculture,
RARS, Vijayapura, Karnataka, India

Dr. G. Lal
Director
ICAR-National Research Centre on
Seed Spices, Ajmer, Rajasthan, India

Dr. M.S. Pendke
Agricultural Engineer, AICRP for
Dryland, VNMKV, Parbhani,
Maharashtra, India

Dr. Anil Kumar Kothari
Chief Scientist & Professor, SWE,
Dryland Farming Research Station,
Bhilwara, Rajasthan, India

Dr. O.P. Pareek
Senior Scientist & Head (Agricultural
Engineering), Krishi Vigyan Kendra,
Bhilwara, Rajasthan, India

Dr. D.H. Ranade
Principal Scientist
Operational Research Project for
Dryland Agriculture, COA, Indore,
Madhya Pradesh, India

Dr. J. K. Balyan
Assistant Professor
Dryland Farming Research Station,
Bhilwara, Rajasthan, India

Dr. Deepak Sharma
Professor
College of Technology and
Engineering, MPUAT, Udaipur,
Rajasthan, India

Dr. Shakir Ali
ICAR-IISWC, Research Centre, Kota, Rajasthan, India

Dr. Dhurendra Singh
Division of Crop Improvement, ICAR-CIAH, Bikaner, Rajasthan, India

Dr. Devendra Pal Saini
Professor (PBG) and Project Incharge, AICRP on Cotton Agriculture Research Station, Banswara, Rajasthan, India

Dr. C. M. Yadav
Scientist (Animal Production) Krishi Vigyan Kendra, Bhilwara Rajasthan, India

Dr. S. M. Kumawat
Prof. Agronomy, Department of Agronomy, S K R A U, Bikaner,`1 Rajasthan, India

Dr. Om Prakash
Principal Scientist, Division of Agri. Extension, ICAR-IARI, New Delhi, India

Dr. R.K. Sharma
Assistant Professor, Dryland Farming Research Station, Bhilwara, Rajasthan, India

Dr. (Mrs.) Prakash Panwar
Assoc. Professor
College of Agriculture, Bhilwara, Rajasthan, India

Contents

Chapter 1

Initiatives for the Growth of Agriculture for Sustainable Development and Empowerment of Farmers

Dr. Narendra Singh Rathore

Deputy Director General, Agricultural Education Division - ICAR, New Delhi

Present Situation of Agriculture in India

Agriculture contribute approximately 14% to India's GDP. Directly or indirectly more than 50% of the total workforce is employed in this sector. Rapid urbanization, growing population and growth of other sectors & their promising employment is impacting the farm productivity. Increase in production, productivity, profitability is main aim of Agriculture sector. Need for quality, remuneration, sustainability and trading in agriculture field. Need to increase in efficiency and per man productivity. Need to increase in the yield of land per unit of area. Improvements in Agricultural Operations, Simplicity, Convenience and Efficiency. Commercial Agriculture i.e. federating farmers in to business group. Skill development in rural living and such allied activities. Need to implement IFS, FS, GAP, AP and AM.

Strategy

1. Improved Work Culture & Good Governance (Decentralization, Empowerment and Accountability).
2. Motivation of Scientists, Staff, and creating state-of-the-art infrastructure.
3. S&T for meeting new challenges i.e. land and water degradation, increasing income & employment opportunity.

4. Promoting Innovation, Teamwork and Entrepreneurship Development.
5. Commercialization, Globalization and Improved Knowledge Management System.
6. Efficient utilization of agricultural inputs for increased productivity, profitability and environmental sustainability.
7. Developing new breed of skilled human resources equipped to meet new challenges of agriculture.
8. Innovating extension education for economic empowerment of the farmers.
9. Making farmers market strategist for enhanced income by processing and value addition, developing models for federating farmers and linking to markets.

Challenges in Agriculture : P Q R S T Formula

To enhance agriculture status of India, we have to face many challenges which inscribed as a PQRST formula.

Productivity : Integrated Gene Management, High yielding varieties tolerant to biotic and abiotic stresses, Cloning of genes/ promoters for economic attributes.

Quality : Good Agricultural Practices for quality enhancement, check on use of chemical insecticides and pesticides.

Remunerability : Reducing cost of production, technological innovations in value chain management and PHM.

Sustainability : To maintain health of our natural resources to develop sustainable agricultural producion systems.

Trading : Farmer to be not only an Efficient Producer but a well Informed Market Strategist.

Research Priorities

- Enhancing the productivity and profitability of field and horticulture crops through technological led innovations.
- Promoting basic and strategic research in new and emerging areas i.e. Genomics, Nano technology, Biotechnology, Food technology,

Renewable Energy, Remote sensing, GIS & GPS, ICT, PHT and Value addition, Precision Agriculture, Genetic Resource Enhancement, NRM, etc.

- Bio-security and plant/animal health management. Resilience to Climate Change and Smart Agriculture. Time saving, Water saving and diseases resistant seed development.

Agricultural Education

Quality assurance should be in education and it should meet stake- holders expectations. It also concentrates on employability, market demands, industrial needs, entrepreneurship and global competi-tiveness. There should be developed programs for meeting new challenges of climate change, increased productivity and profitability with increasing the efficiency of input use, processing and value addition, quality of produce to meet international standards. It should enhance strong Industry-Institute Partnership for global competitiveness.

Higher Agricultural Education

In Higher Agricultural Education attracting talented students programs, competent faculty and state of the art infrastructure needed. There should be quality human resource development. It should maintain global standards & enhance competitiveness. There is Student READY Programme and Experiential Learning.

In Plant Training/ Industrial attachment. Rural Agriculture Work Experience.

Technology Transfer and Extension

Developing new models of effective technology transfer to farmers for bridging the wide gap between the potential and realized productivity. Ensuring farmer first approach in evolving effective mechanism for technology transfer and adoption.

KVK to work as centre of TADA-CB.Use of ICT tools for Technological empowerment of farmers. Helping farmers to be market strategist and Market research for demand projection, enhancing income through technology intervention.

Mission mode approach on soil testing, fertility mapping and rain-water harvesting. More production with efficient use of water, more area under rainfall cultivation. Climate Smart Agriculture (SHAPE). Skill Development and Retaining Youth in Agriculture. Farmer led Innovations with Natural Farming (UBA). Awareness on social issues like GM Crops, Climate change, Energy and Water Management, etc. Farmer's markets, producer's organizations, joint farming groups of landless. Farmers to play a key role in 'Digital India', 'Skill India', 'Swach Bharat' and 'Make in India' program.

Infrastructure Development

For Precision Agriculture, Secondary Agriculture, Conservation Agriculture, Speciality Agriculture, Vertical Agriculture, Canopy under Canopy, Aero phonic and Hydro phonic, etc.

Demonstrate use of different models i.e. Organic farming, WRRRS, IFS, Climate change study centre, GIS and remote sensing centre, RE centre, Learning resource centre, Processing and value addition centre, Information technology centre, etc.

Farmer First

It aims to move beyond production and productivity and to recognize the complex, diverse and risk prone realities of majority of the farmers. Enhance farmers-scientists contact with multistake-holders' participation. It aims at enriching farmers-scientists interface for technology development and application. It is achieved with focus on innovations; feedback; multiple stakeholders participation, multi method approaches, vulnerability and livelihood interventions.

Mera Gaon-Mera Gaurav:

To enhance the direct interface of scientists with the farmers, which will hasten the lab to land approach.

To provide farmers with required information, knowledge and advisories on regular basis. Groups of scientists will select villages and will remain in touch with that village and provide information to farmers on technical and other related aspects in a time frame through personal visits or through telecommunication.

In this way, 20,000 scientists of National Agricultural Research and Education System (NARES) can work directly in villages.

Attracting and Retaining Youth in Agriculture

Realizing the importance of rural youth in agricultural development especially from the point of view of food security.

To attract and empower the Youth in Rural Areas to take up various Agriculture, allied and service sector enterprises for sustainable income and gainful employment in selected districts. To enable the Farm Youth to establish network groups to take up resource and capital intensive activities for secondary agriculture.

To demonstrate functional linkage with different institutions and stakeholders for convergence of opportunities available under various schemes/program for youth.

Student Ready

The term READY refers to "Rural and Entrepreneurship Awareness Development Yojana". Student READY is Skill development initiative to strengthen students with skills. To enable them to tackle global challenges, and to improve both their employability as well as ability to set up a venture. Student READY concept signifies this as a Finishing school for the undergraduate students.

Expected Outcome

Leadership in Agriculture Research and Education at international level. Effective contribution towards ever green revolution in Agriculture. Enhancing farm production, productivity and profitability coupled with environmental sustainability. Increased efficiency in the use of inputs & thus, enhanced farm income, employment generation, entrepreneurship development and environmental protection. High quality research through state of art infrastructure, attracting talented faculty with improved governance.

❑❑❑

Chapter 2

Techniques for Management of Arable and Non-arable Lands to Enhance the Productivity in Rainfed Areas

J.K. Balyan, A.K. Kothari and Devyani Dodia

Dryland Farming Research Station, (Maharana Pratap University of Agriculture & Technology, Udaipur), P.B.No. 62, Arjia, Bhilwara (Rajasthan)-311001

Introduction

Rainfed agriculture occupies 68% of India's cultivated area and supports 40% of the human and 60% of the livestock population. Thus, rainfed agriculture has a crucial role to play in sustaining the economy and food security. However, aberrant behaviour of monsoon rainfall, eroded and degraded soils with multiple nutrient and water deficiencies, declining ground water table and poor resource base of the farmers are principle constraints for low and unstable yields in rainfed areas (Singh, et al., 2004). Increasing crop productivity to meet food requirements of teaming millions in our country poses a greater challenge. In this context, there is a need to enhance the productivity of rainfed crops from at least 1-2 t to meet the food requirements by 2020 AD. Hence, the situation calls for efforts to intensify the production in both, time & space. This could be possible by developing appropriate cropping and farming systems in rainfed agriculture.

Bhilwara district is situated in agro climatic zone IV-A of Rajasthan. This district has total geographical area of about 10.46 lakh hectare, out of which the cultivated area is about 4.34 lakh hectare and at present about 57% of the net sown area is rainfed which contributing 40% of the total food production, supports 40% of human and 2/3rd of livestock population. In the zone IVA of Rajasthan, there are 6.77 lakh farmers family, out of which 73.43 % are marginal and small in Bhilwara district, there are 3.2 lakh farmers family, out of which 71.86% are marginal and small with poor resources. Management Techniques of arable and non-arable land will be helpful to increase yield per unit area per unit time, profitability, sustainability and provides balanced nutritious food for the farmers in rainfed areas. It also creates pollution free environment and provide opportunity for effective recycling of one product as input to other component, money round the year and solve the energy, fodder, fuel and timber crisis, avoids degradation of forests and enhance the employment generation, increase input use efficiency and finally improve the livelihood of the farming community. Small and marginal farmers with capital scarcity, risk avoidance objectives, and a cautious learning process rarely make drastic changes in their farming system. Rather, they proceed in a step-wise manner to adopt one and sometimes two new inputs or practices at a time.

1. Land capability classification

Table 1: Capability class with suitable farming system

Capability class	Suitable farming system
i)	Crop production & intensive cultivation
ii) & iii)	Moderately intensive cultivation, Alley cropping, Agro-horticulture
iv)	Restricted cultivation- lay farming, Agro-horti system
v) & vi)	Horti- silvi-partoral system
vii)	Low density, long rotation, tree plantation
viii)	Recreation wildlife, watershed protection

New scheme for Land capability classification : Shang (1990) presented a treatment oriented of new scheme for Land capability classification for hilly western areas:

Table 2: Symbols for most intensive tillage or uses

Slope/Soil depth	Gently sloping $<7^0$	Moderately sloping 7^0-15^0	Strongly sloping 15^0-20^0	Vary strongly sloping 20^0-25^0	Steep 25^0-30^0	Very steep $>30^0$
Deep (D) >90cm	C_1	C_2	C_3	C_4	FT	F
Moderately deep (MD) 50-90cm	C_1	C_2	C_3	C_4/P	FT/F	F

Shallow (S) 20-50 cm	C_1	C_2/P	C_3/P	P	F	F
Very Shallow (VS) <20cm	C_1/P	P	P	P	F	F

Symbols for most intensive tillage or uses

C_1 Cultivable and 1, upto 7° slope, requiring no or few intensive conservation measures, e.g. contour cultivation, strip cropping, vegetative barriers, rock barriers and in larger farms broad base terraces.

C_2 Cultivable and 2, on slopes between 7° to 15° , with moderately deep soils needing more intensive conservation, e.g. terracing, hexagon, convertible terracing for the convenience of four-wheel tractor farming. The conservation treatments can be done by medium sized machines, such as bulldozer.

C_3 Cultivable and 3, 15° to 20°, needing bench terracing, hexagon and convertible terracing on deep soil and hillside ditching, individual basin on less deep soils. Mechanization is limited to small tractor or walking tractor because of the steepness of the slope. Terracing can be done by a smaller tractor with 2.5 m wide blade.

C_4 Cultivable and 4, 20° to 25°, all the necessary treatments are likely to be dose by manual labour. Cultivation is to be practiced walking tractor and had labour.

P Pasture land, improved and managed pasture. Where the slope is approaching 25^0, and when the land is too wet, zero grazing should be practiced. Rotational grazing is recommended for all kinds of slopes.

FT For good trees of fruit trees. On slopes of 20° to 25°, orchard terracing is the main treatment supplemented with contour planting, diversion ditching and mulching. Because of steepness of the slopes, interspaces should be kept in permanent grass cover.

F Forest land, slopes from 25° to 30° or over 30° , where the soil is too shallow for any of the above mentioned conservation structures.

i) Slopes and number of slope classes can be modified to meet contry's needs and some F lands between 25^0 to 30^0 can be used for agro-forestry purpose.

ii) Any land which is too wet and occasionally flooded or too stony, which prevents tillage and treatment should be classified as: (a) below 25^0: pasture (b) above 25^0: forest

iii) Gully dissected lands which prevent normal tillage activities: forest (over 25^0) or pasture (below 25^0)

a) Slope classification: Slopes are divided into six categories, each having its implications for conservation treatments and the kind of tools to be used:

i) **Slopes <7^0** : Flat to gently sloping. broad base terraces of other simple conservation treatments can be used upto 7^0, Full mechanization for cultivation is applicable in this category. This slope class may not be common in hilly watersheds.

ii) **Slopes 7^0 to 15^0** : Moderately sloping. Medium-sized machines, such as a bulldozer can be employed for bench terracing. Four wheel tractor use for mechanization and cultivation can be applied.

iii) **Slopes 15^0 to 20^0** : Strongly sloping. Small sized machines can be employed for conservation treatments. Small tractors. or walking tractors can be used for cultivation.

iv) **Slopes 20^0 to 25^0** : Very Strongly slope land manualy for building the structures. Hand labour and walking tractor for cultivation.

v) **Slopes 25^0 to 30^0** : Steep slope land only for permanent tree crops, such as food trees, fruits, forest or agroforestry.

vi) Slopes >30 : Very steep. Needs forest cover.

b) Soil depth

Soil depth is divided into four classes. Here, the depth refers to the effective depth of the soil which machine or manual labour can cut for conservation treatments and which plant roots can penetrate:

i) **Shallow (20-25 cm)** : Only below 20% slopes can this be cultivated with conservation treatments.

ii) **Moderately deep (50-90 cm)** : On a 25% slopes for instance, it needs about 76 cm of soil to make narrow terraces of 2.5 m wide.

iii) **Deep (>90 cm)** : No further soil depth classification is needed because the riser or terrace is limited to 1.8 m height, which is 0.9 m cut and 0.9 m fill.

c) Other limiting factors

Land which is too wet, has poor drainage, occasionally floods or is too stony, which permanently limits the tillage or treatment, should be classified for lower or less intensive uses. On slopes under 25% such land can be used as pasture, whereas on slopes over 25% forest cover is proper as far as erosion control is concerned. Gully-dissected land which prevents any tillage activity should be put under permanent cover.

d) Soil conservation treatment

In addition of the most popular conservation treatments on gentle slopes (below 7^0), such as broadbase terraces and strip cropping, etc., six major treatments for steeper slopes are taken into account for the basis of the new classification scheme. These six treatments are particularly suited for the humid tropics. Bench terraces, hillside ditches and individual basins can be used to treat slopes up to 25^0 to 30^0 slope. Convertible terracing and hexagons for full mechanization are to be employed on slopes upto 20^0. All of them are mainly reverse sloped terraces of varying widths. Later another type of terracing was added: Intermittent terraces.

Soil conservation is the preservation of soil from deterioration and loss by using it within its capabilities and applies the conservation practices needed for its protection and improvement.

I. Management practices for arable lands

i) Reduction of rainfall impact

ii) Reduction of erosion

iii) Control of runoff and conserving runoff.

Measures for *in-situ* moisture conservation

- Deep summer tillage
- Field bunding
- Contour cultivation
- Compartmental bunding
- Contour bunding

- Inter cropping system
- Broad Bed Ridge Furrow System (BBRFS)
- Ridge and furrow system
- Strip cropping

These are depending upon the degree of erosion hazards. Agriculture lands are treated with suitable soil conservation measure to protect it from degradation & to preserve the top fertile soil. The selection design of soil and water conservation for agricultural land usually depends on the soil types, land slope, soil depth & rainfall of the area. The measures adopted are broadly classified in to two groups:

A. Agronomical measures

B. Engineering measures

A. **Agronomical Measures** : These adopted when the land slope is less than 2% and erosion problems are not severe vegetative measures can be adopt in combination with engineering measures. Agronomical and vegetative methods for controlling water erosion consist of:

a) Contour cultivation

b) Contour vegetative hedge

c) Contour vegetative bund

d) Strip cropping

a) *Contour cultivation :* Contour cultivation is the system of farming on sloppy lands in such a way that ploughing, preparing land, planting sowing & intercultural operations are done on the contours. It creates numerous ridges & furrows which have to cheek soil loss in humid areas and conserved soil and moisture in low rainfall areas.

b) *Contour vegetative hedge :* Vegetative barriers on contour with live bunds alone or with vegetative hedge are to be useful in preventing erosion & conservation of moisture in soil, when placed at suitable intervals.

Functions

- To break the slope length, reduce runoff velocity & increase infiltration opportunity.

- To reduce erosivity and transpiration capacity of the surface runoff.
- To Carry out the cultivation & planting operation on the contour for increasing the soil productivity.
- To stabilize contour cultivated areas on steeper slope e.g. *Saccharum munja, Lacerius sindicus* (Sevan grass) & other local spp., *Vetiver jejionodes* (Khas grass) & lamon grass can be used for semi-arid areas.

Benefits

i) Cheaper to make, take less space are self-maintaining established.

ii) Contour vegetative hedge avoid problem of water-logging.

iii) Fertile top soil is loss in earthen bund and a wide stream of cultivated land over the entire length of bund, due to earthen bund it become out of cultivation.

iv) Does not require engineering skill.

v) No need of providing any waste weirs for safe disposal of excess water.

Limitations : Adaptability of contour hedge is very poor.

c) *Contour vegetative Bund :* It is constructed in the arable land which have slope range of 2-6%. The recommended cross section of contour vegetative bund is $0.34m^3$ in the 600 mm per annum rainfall areas. The bunds constructed across the slope & inside ditch are executed, where the vegetative hedge are painted.

d) *Strip Cropping :* Growing of alternate strips of erosion permitting crops, such as maize, sorghum, bajra, cotton, etc. and erosion resisting crops (Blackgram, greengram, mothbean & groundnut etc.)

- The width of strip of erosion resisting & erosion permitting crops depends on slope, soil texture, rainfall characteristics and type of crops.
- Generally the steeper the slope the greater is width of erosion resisting crop (ERC) & smaller the width of ERC.

i) **Contour strip cropping** : Layout, tillage operation and planting of crops are practiced along the contour. The crops follow a definite rotational sequence in contour strip cropping. Generally, the width of strip may vary and maximum upto 200 feet of a strip. The strip cropping is followed where slope is upto 5% and land capability class II & III type.

ii) **Buffer strip cropping** : Growing the strip of grass or legume crop, that completely covers the ground. These strops are placed in between contour strips. They are planted in regular rotation. These are followed upto 5% slope. The width of Buffer strip cropping generally 8-20 feet.

iii) **Field strip cropping** : Under irregular topographic condition, the width of two contour lines very too much. In these situations contour strip cropping is not practicable. The strips of crops are uniform width & placed across the general slop.

iv) **Border strip cropping** : At both the edges, strips of permanent grasses or crops are grown & these strips are grown against the main strip. These are used to check or restrict the erosion.

v) **Correction strips** : Make correction strip where width is not equal in contour strip.

Table 3: Width of erosion permitting and erosion resisting crop strips as per land slope

Slope	Width of EPC (m)	Width of ERC (m)
1%	50	10
2%	30	6
3%	15	5

5. **Crop rotation** : Erosion resisting crops are very necessary in crop rotation, grasses, small millets, pulses crop effectively check the soil erosion & maintain the soil fertility. In a crop rotation, growing of cereals & pulses in alternate way or grown as mixed crop or intercrop.

Advantages of crop rotation

i) Help in maintain & improve the soil fertility.

ii) Protracts the soil borne disease and already grown weeds.

iii) Check soil erosion

iv) It provides employment to farmer whole year.

v) Crop utilize the moisture from different depth due to different root length.

vi) Soil moisture can be conserving in one season & can be used in next season.

vii) Can get whole year production in crop rotation.

viii) It helps to maintain good level of organic matter and nitrogen in soil

ix) Fulfill all needs of farmer in respect to cereals, pulses soil sod.

x) Insect, disease & weed can be controlled by using ideal crop rotation.

xi) Farmer can use all resources by using crop rotation.

6. Tillage

Advantages of tillage

i) Improves the soil tilt.

ii) To destruct the stalk roots of old crop, weeds and other materials like stables of stem and roots.

iii) Good air circulation for roots

iv) Provides good seed bed

v) Lowers down the quantity of runoff water.

vi) Destroys the insect & organism of disease.

vii) Mixes the green manure of FYM and other manure in soil.

viii) Checks the soil erosion due deep ploughing.

ix) Helps to soil for check the extra water and air.

x) Provides a good medium to grow & establishment of plants.

7. **Mulching** : Mulching is process of covering soil surface with crop residues, straw, leaves, sand, plastic or paper with the object & protecting soil from the impact of rainfall and to check the soil erosion. It is also helpful to conserve the soil moisture.

8. **Cover crops** : Those crops which grow quickly and cover the soil surface called cover crops. These help to check soil erosion & reduce the runoff and also improve the soil fertility.

9. **Use of manures** : Manures have capacity to tie the soil particles and it checks soil erosion, manures also increase the water holding capacity of soil.

10. **Biofertilizers** : The micro-organism are used in soil by three ways, seed treatment, soil treatment and dipping of roots of transplanting crop for fixing the atmospheric nitrogen plant roots or make availability of N&P in the soil & plant. These are also helpful for plant growth & development due to creation of plant growth hormones. Therefore, use of biofertilizers reduce the cost of cultivation & increase the yield with maintaing good soil health. INM reduces the soil deterioration in watershed areas.

11. **Seed rate and right method of sowing** : Soil erosion can be checked by sowing the seeds in row against the slope and increased in seed rate help quickly cover the soil surface.

12. **Methods of irrigation** : Soil erosion can be checked by using ridge and furrow method of irrigation. Sowing of crops on contour and irrigation in these contour furrow is helpful to chech the soil erosion.

13. **Methods of harvesting of crops** : At the time of harvesting crop, stubbles are let in the field, these also help in check the soil erosion.

B. Engineering Measures : Engineering measures can play a vital role in erosion control on agricultural land. They supplement with agronomic practices when the latter alone are not adequate. These measures are generally adopted for the slope of land more than 2% and engineering measures applied for overall treatment of the land, water disposal and make the water harvesting structure. Mainly engineering measures are used in watershed areas as follows:

1. Overall area treatments.
2. Water disposal treatments.
3. Water harvesting structures
4. Erosion control structures

Principles

1. Increasing the time that water stay where it falls, thereby allowing more runoff water to be absorbed and held in the profile.
2. Breaking up a long slope in several short one.
3. Reduce the speed of runoff to loss than the speed which cause soil erosion.
4. Protection of drainage channels against damage of excessive runoff.

Table 4: Simple criteria for deciding needed measures as per Land capability class

Land capability class	Measures / practices	Advantages	Disadvantages
II	Contour cultivation	i) In situ moisture conservation increase ii) Crop yield increase iii) Reduce runoff & prevent soil erosion	Nil
III	i) Contour/ Graded bunding	Land slopes are broken down in to more level compartments, conserve moisture prevent soil erosion	Nil
	ii) Vegetative barrier contour cultivation strip cropping	Conserves moisture & reduce soil erosion	
IV	Stone wall Puerto/ Rico terraces	Sloping land with limited depth of soil can safely be put under cultivation check soil erosion. In due course of time land become level	Feasible only when stone are available locally cost is more

Engineering measures for soil and water conservation for arable lands:

i) Contour bunds

ii) Field bunding

iii) Waste wires

iv) Graded bunds

v) Terraces

vi) Trenching

vii) Diversion channel

i) **Contour bunds :** The main criteria for construction of bund on contour to intercept the water before it attains erosive velocity break the long slope into several short ones and allow more runoff for infiltration & held by the soil. These bunds are constructed on contour and height of the bund is determined on the basis of amount of water to be intercepted by bund.

ii) **Field bunding :** Periferal bunding or field bundig is helpful to conserve the rain water in the cultivated field and reduce the erosion and conserve the moisture.

iii) **Waste weirs :** Waste wires constructed for safe disposal of water, these are the outlets which permit the safe disposal of excessive runoff water. The waste weirs constructed in bunds to protect the bund.

Functions

i) To prevent damage to the standing crop from excessive submergence.

ii) To protect the contour bund from damage by excessive heavy runoff.

Types of outlets

i) Clear overall stone wires

ii) Channel wires

iii) Ripe outlet

iv) **Graded bund :** Graded bunds or graded terraces are laid along a predetermined longitudinal grade instead of along the contour.

Graded bunds are constructed in medium to high rainfall areas having annual rainfall more than 600 mm and soils are poor permeable.

v) **Diversion channel :** It is a channel with a supporting ridge on the lower side. These are constructed lower side of ridge across the slope for the purpose of intercepting surface runoff and conveying it to natural stream with the sage outlet.

vi) **Terraces :** A terrace is an earth embankment or a ridge and channel constructed across the slope of the land at an appropriate location to control runoff and minimize soil erosion. Terrace occupying a slope or discharging runoff into 1 or more outlets called terrace system. Following types of terrace are constructed:

a) **Bench terrace :** These are constructed on steep slope land areas and construction of steep like field along the contours by half cutting & half filling of slope. Original slope of land is constructed into level fields and all hazards of erosion are eliminated. It is adopted, where the soil slopes are > 15%

- **Levelled terraces :** These terraces which are levelled from back to front & water flow on these terraces is also equal. These are followed where deep soils are present.
- **Slope outward :** These are followed where rainfall & depth of land is less.
- **Slope Inward :** These are followed where rainfall and slope is more.

b) **Graded channel terrace or runoff interception terrace :** These are followed where slope is 10-20%. These terraces are made with slopes drain the water outside so that the water flow & soil erosion can be lower down.

c) **Absorption type terrace or level terrace :** These terraces are made where less rainfall occur and soil which absorb water quickly and here are no damage to crop even in the absence of water drainage. These are made where slop is less than 3%.

d) **Puerto-rico terraces or stone wall terrace :** In this type of terrace, a mechanical barrier of 30-45 cm height is prepared or at desired distance vegetative barriers are grown on contour from upside to downside and ploughing is done towards the slope so that more soil come with runoff water. These terraces are made up of stones in vallies.

e) **Zingg terrace :** It is suitable where slope is 3-10% land is divided into 2 pieces one is done which will not levelled and another is receiver which will be levelled. Ratio of donor and receiver should be 2:3 and 0.3 to 1.5 m wide ridge made at the end after the 2:3 ratio to check the water.

II. Management practices for non-arable lands as per land use capability classification

1. Contour trenching

Contour trenching implies excavating trench along with the contour or along a uniform level across the slope of the land. Bunds are constructed in downstream along the trench with material excavated out of them. These trench break the slope length & reduce the velocity of runoff and retorted its:

i) **Continuous contour trenches :** These trenches store more water but they are more expensive more over these are chance of water flowing along the upper edge if trenches are breaks. These trenches are essentially used for conservation of moisture in low rainfall areas and need careful layout.

ii) **Interrupted contour trenches :** These intermittent trenches are suited to high rainfall areas. These may be constructed on contour line or staggered with respect to each other.

2. Trenches

a) **Contour trenches :** These trenches are made of equal depth on the contour. These are made on slope hills, barren and vacated land and forest lands for moisture conservation. Bunds are made downward on slopy side by soil of trenches. Contour trenches lower down the flow of runoff water. Contour trenching is a practice of excavating trenches along a uniform level across the slope of the land.

b) **Gradonies :** Gradonies are steepy inwards sloping narrow bench terrace constructed on contour, these are suitable for uniformly steep slopy lands.

c) **Box cum pit trenches :** In this system box type trenches with the size 1.5 m X 0.3 m are excavated. Box to box distance is kept 2.5 cm and spacing is maintained 4.0 m from centre to centre.

Contour lines are based row of the boxes. The excavated soil is heaped on down slope to from a berm.

d) **Staggered trenches :** These trenches are made on sloppy and un- levelled land to check soil erosion and to make more effective the silvipasture system. These trenches are excavated on contour lines and are small in length.

e) **Fengo** (Sum trench) : The trenches are made on sloppy and unlevelled land. Trenches also excavated on contour but the excavated soil is heaped on upper slope so that excavated soil particle drain in trench & it becomes a levelled terrace. In these trenches, grasses can be grown easily.

3. **Gradonies :** Gradonies are steeply inward sloping narrow bench terrace constructed on contour, usually gradonies are suitable for afforestation in uniformly steep slope land. Based on the steepness of slope, vertical interval is kept of gradonies and the width of gradonies also varies. The material dug from the inner side is heaped on the outer edge in order to make a bench with an inward slope. In the middle of the gradonies, pit are constructed. On an average 1000 trees are planted within the gradonies & interspaces in one hector area. Gradonies were proved to be quite successful in *Arawali hills* of southern Rajasthan for afforestation.

4. **Gully control measures :** Formation of gully takes place due to improper use of land for a long time. Once a gully is formed, the cost of reclamation is so high and the reclamation can hardly justified in simple economic term for control the gullies formation measures. Gullies can be formed with following aims.

i) **Planning for control of runoff :** The proper planning should be followed for control of runoff from watershed is essential, for this adopted the following measures:

a) **Retention of runoff :** To control gully formation it is advisable to retain the maximum possible runoff in the catchment, for this common measure like bunding, terracing, contour farming, strip cropping, mulch tillage, ridge planting, etc. can be used.

b) **Diversion of runoff :** The excess runoff water should be safely diverted away gully through the graded terrace and field diversion channel can be used to convey the water into a vegetative water ways.

c) **Conveying runoff through gullies :** If the gully is located in a natural depression and the rainfall & soil condition are such that the excess runoff can not be stored in the watershed, then mechanical structure should be constructed at suitable location and sod can be transplanted at critical location.

ii) **Vegetative measures for gully control :** The vegetative may consist of grass cover & for this purpose various types of plants, the measures are adopted as follows:

a) **Sod flumes :** For a small watershed the gully can be treated with flumes through which water is allowed to flow for handing medium runoff for handing large size sods, for low runoff strip sod is adequate.

b) **Sod check :** Two types of soil checks are used:

(i) Sod strip (ii) Sodded earth fill

Objectives of sod check

i) Sods are places in strip along the gully at a regular interval.

ii) To stabilize small & medium size of gullies sodded earth fill may be used.

c) **Shrubs checks :** Shrubs or tree branches can be used for stabilization of small gullies having gentle slope with limited drainage area. Trenches are made across the gully slope and shrubs firmly placed in them. Shrubs check retained the velocity of eater & trap the soil particles.

d) **Tree and shrubs :** Plantation of trees and shrubs are useful for gully control. Appropriate species of shrubs and trees should be planted at optimum moisture content available in gullies.

iii) **Temporary gully control structure :** Under this system, mechanical measures should be adopted to check washing away of vegetation by the large volume of runoff. Temporary structures are made of bush, wood rocks, woven wire, sod or earth.

a) **Bush dams** : Bush dams are suitable to stabilize small gullies, they are cheap and easy to construct with brush and hay. They do not require any skilled lab our and can make locally available materials.

b) **Loose stone dam** : Loose stone dam are temporary structure but have much longer life in comparison to bush dam. Woven wire netting may be used to provide stability to these dams.

c) **Plant or slab dams** : These are mainly constructed in north eastern region of the country. They need less labour and may also be used for relatively large drainage areas.

d) **Log dams** : Log dams are used in those areas where timber is available in abundance. The construction of log dam is almost similar to plank dam.

e) **Gabion check dam** : Gabion check dam are used in areas where stone are available in abundance. Usually, a gabion is a rectangular shaped cage which made of galvanized wire and filled with locally available stones.

iv) **Permanent Gully control structure :** Permanent gully control structure stop the advance of gully heads, stabilize its grade & assist for establishment of vegetation. They are constructed masonry through cement stone, earth, etc. Usually these are expensive.

v) **Sediment retention structure :** Sediment retention structures are used to trap the eroded soil before it can reach a water body. These structures should be employed only as a back system.

The working principle of sediment retention structure is based on slowing the runoff velocity & retuning the suspended soil particle settle by gravity. These structure may be inclusive straw, pole dykes, filter, sediment traps in which soil particle settle through when water flow through the influence of gravity.

vi) **Retaining walls :** Retaining walls are constructed in order to stabilize hill slopes. They are classified into two categories:

a) Retaining walls subjected to water pressure

b) Retaining walls subjected to earth pressure

vii) **Reclamation of Ravines lands :** Ravines are network of gullies. The ravines are an indication of very bad management & land resources in the country. Small ravines are reclaimed by clearing minor leveling & construction of check bunds at horizontal intervals. Control of Ravines include:

i) Closure of biotic interference

ii) Construction of contour and peripheral bunds

iii) Protection of Ravines by plugging

Alternative landuse system in watershed area : Alternative land use system can be defined as a perineal system adopted to replace or modify the traditional landuses. Alternate landuse has more potentiality and flexibility of landuse than the traditional crop production system.

Alternate landuse system aim is a suitable farming system with machining the land capability class like -Tree farming, lay farming, pasture management, etc. The main aim of the alternate land use system is to generate assured income with minimum risk through efficient utilization of available resources.

Table 5: Alternate landuse system as per landuse capability classes

FoodArable land (II&III)	Fodder(IV & V)	Fuel/Timber/Fiber(VI & VIII)
Alley cropping	Hortipastoral	Tree farming
Agro-horiticulture	Silvipasture	Timber cum fiber
Inter cropping	Lay farming, Pasture Management	

II **Farm Planing & Allocation of resources :** Food, fodder, fuel, timber are the primary requirement of the farmers. Improved alternate landuse system not only meets the various requirements but also generate continuous and stable income.

III **Need base alternate landuse system :** Among the several needs of a farmer, food always remains the first priority item and after food, the fodder is important requirement of farmer. Among the several needs of farmer, food always remains the first priority than others.

A. Alternate land use system for arable land

i) **Alley cropping :** It is also known as hedge row intercropping, it involves raising of fodder crops in alleys from by hedge rows of tree or shrubs. Alley cropping is a suitable alternative to shifting cultivation.

Advantages

i) It provides higher total biomass per unit area than arable crop alone.

ii) Utilizes of season precipitation

iii) Provides green fodder during lean period.

iv) Provides additional employment during off-season.

v) When planted along contours on slopy lands, it provides a barrier to runoff, holds the silt & conserve the moisture.

ii) **Agroforestry system :** Agroforesty system which involves integration of trees with crops. When growing of field crops between the inter row space of tree plants. Trees are used with a top food tree/fuel purpose is called agroforestry system.

iii) **Agri-silvicultire :** Intercropping with NFTs (Nitrogen fixing trees): raising of NFTs in arable land not only provides rich organic matter and atmospheric nitrogen but also improves soil structure and prevents land degradation. Most of the common Nitrogen fixing trees are grown generally: *Accacia albeda, Glyrecidia sp., Parkin sonia, Pongamia glabra, Prosopis cenceraria, Leueiane leucocifela, Sesbania,* etc.

iii) **Agro-horticulture system :** This system is also termed as food cum fruit system in which short duration arable crops are raised in the interspace in fruit trees, fruit trees: Aonla, Ber, Phalsa, Pomegranate, Anarete.

B. Alternative landuse system for Non- arable lands:

i) **Horti-silvi-pastoral system :** Class IV and above classes land are termed as non-arable lands. These are unfit for crop production. The topography of these lands are highly undulating. These are lack vegetative cover, runoff or soil & water is dominant. These lands have to be brought under close canopy vegetative cover to prevent these lands for further degradation. Grasses and legumes are associated with hardy fruit trees/Top food trees should be planted on non arable lands.

ii) **Horti-pastoral system :** Defined as one of the agroforestry system which involves integration of fruit trees with pasture. When a fruit tree is replaced with a top food tree is called silvipastural system.

Selection of fruit trees : The fruit tree species should be so chosen they completely or less competitive for under stray plants. Slow growing and late bearing fruit tree species like: Jamun, Aonla are highly suitable in a continuous growing system. Custard apple is an ideal option, as it is not used by the live stock and it is hardy fruit plant. In water-logging area and alkali soil, aonla is best suited fruit plant.

Selection of top fodder tree species

i) These tree sp. should be fast growing and it should be have multipurpose uses.

ii) High portability and digestibility of forage

iii) Good coppicing capacity

iv) Ability to withstand brushing, transplanting and tolerant to drought and extremes of the climatic conditions.

Table 6: Selection of suitable grass and legume crops species

Pasture Spp.	Minimum rainfall	Soil type	Seed rate	Yield
Sevan grass (*Lasariyus cindicus*)	250 mm	light	3-5 kg/ha	3-5 t/ha
Buffer grass	300 mm	Versalite	5-6 kg/ha	4 t/ha
Dhaman/Anjan grass (Black Dhaman) *Cenchrus setigerus/ Cenchrus ciliaris*	250 mm	Versalite	8-10 kg/ha	3 t/ha
Blue panic	500 mm	Versalite	3-4 kg/ha	3 t/ha
Marble grass (*Dicanthum annulatam*)	375 mm	Medium to heavy	4-5 kg/ha	2-5 t/ha
Napier grass/ Elephant grass (Pennisatium perpariam)	750 mm	Medium to heavy	5-7 kg/ha	3-5 t/ha
Legume perennial				
Stylosanthus hamata/ Stylosanthus scabra	450 mm	light to medium	5-7 kg/ha	3-5 t/ha
Cowpea	300 mm	– –"– –	30-40 kg/ha	3 t/ha
Guar	300 mm	– –"– –	25-30 kg/ha	3 t/ha
Horse gram	300 mm	– –"– –	40 kg/ha	2 t/ha

Establishment technique of grasses : Good seed bed beginning of rainy season on a drizzling day and sod should be sown, shallow depth is required. A seed rate of 4-6 kg with soil or sand in two times of its volume would be ensure for good establishment.

Horti-pastoral system : In a horti-pastoral system fruit trees like guava, custard apple were planted and seed of *Stylo* grass & *Cenchrus* were broadcasted one year later between the row spaces of fruit trees to ensure the good establishment of fruit trees.

Silviparstoral system

i) Lay farming

Pasture management : The important technical interventions are identification of suitable grass & legume spp. using suitable establishment technique, fertilization of trees.

Tree farming : It is said that agriculture is for plains having assured rainfall, while trees are for hills, degraded lands and drylands, where greater risk of crop failure exists. Tree can furnish a yield absolutely in many places, where arable crops fail to grow.

Tree raising practices : Boundary planting, live fences and hedges, shelter belt/wind breaks, high density planting tree and shrubs for stabilizing soil & water conservation structures, tree for fedder or vegetative barriers, establishment of apiculture, lac culture, sericulture and tree used for land reclamation.

Tree establishment technique : Survival & growth of trees seedlings depend on nursery and establishment technique. These includes good site preparation, size of seedling, age of seedling, pit dimension, pit filling mixture, amendment, time of planting, introduction of leguminous seedling with suitable rhizobium , suitable input for marginal lands.

Table 6 : Suitable tree species under different rainfall areas

<500 mm rainfall	500-750 mm rainfall	> 750 mm rainfall
Acacia nilolica	*Acacia nitltica*	*Acacia nililica*
Acacia tortilis	*Albizra leback*	*Albizia leback*
Acacia albida	*Azadiracheta indica*	*Azadirachate indica*
Prosopis cineraria	*casurena equisitifloia*	*casurena equisitifolia*
Prosopis juliflora	*Delberagia sisso*	*Delbergia sisso*
-	*Leucaena leucoephala*	
	Tamarindus indica	

Rainwater harvesting

Objectives : Objectives of constructing water storage structures in watershed areas as under:

i) To intercept runoff and moderate peak of low & volume.

ii) To increase more absorption of available rainwater leading to all around improvement of moisture conditions.

iii) To enhance ground water recharge.

iv) To intercept & reduce silt and thus, reduce sediment production in the stream flows and finally in the reservoir.

v) To provide irrigation potential to mini command area both, in *rabi* and *kharif.*

vi) To help in leveling up undulating or eroded area.

vii) To help in reclaiming additional productive land.

viii) To provide remunerative single or double crop & regular job to the people.

ix) To provide employment through the plantation.

x) To increase total production as well as total resources for the entire area.

Function of water storage structure : Storage structure helps in greater absorption of water on watershed areas and also held in between dry spells. These structures assist in moderating water stress in dry spells and help for crop production which is turn works as lasting production against extremes dryness. Data required for planning & design of water storage structure:

a) General Data

i) Location Map : Location of particular existing location of roads and other public utilities, etc.

ii) Hydraulic data : Stream flow records including volume & rates, water requirement percolation evaporation losses, etc.

iii) Climatic data : Maximum, minimum and mean temperature, rainfall storm & intensities.

iv) Geological data : Geological report comprising formation, exposed ground, deposition of permeable nature contributing to serious reservoir leakage.

b) Reservoir data

i) Reservoir map

ii) Road and public utility surveys

iii) Miscellaneous data

c) Data for Dams

i) Dam site map

ii) Foundation exploration

iii) Material exploration

iv) Tail water data

v) Local conditions controlling design of the dam

vi) Local conditions affecting construction.

Earth embankment : Earth embankment used as protective structure in the watershed. The design of earth embankment that is effective & safe in based on sound engineering & construction principle.

General Requirements for Earth embankment

i) Topographic condition of dam site must allow economic construction

ii) Soil material must be available to a possible stable impervious full.

iii) All storage embankments must have adequate mechanical and flowed spillway & facilitate to maintain a uniform water depth.

iv) Large storage embankment should be equipped with a bottom drain pipe, to facilitate maintenance and management.

v) Adequate safety, equipment must be provided drop inlet structure.

vi) All design specification must be adhered to during construction & a sound programme of maintenance must be followed.

Farm ponds : Farm pond is small tanks or reservoirs constructed for the purpose of storage water from surface runoff. Farm ponds are used for providing supplement irrigation to the crops, supply of water to livestock's, fish production, etc. Thus, a farm pond has a significant role in the area of rainfed agriculture.

1. **Dugout farm pond :** Dugout farm ponds are excavated at the site on the soil obtained by the excavation formed as embankment the pond. These ponds are generally constructed in the area having that topography. It is advantages in the areas where evaporation losses are high because it can be constructed to expose a minimum water surface area in proportion to volume.
2. **Embankment farm pond :** These types of pond consist of an earthen embankment or earthen dams which is partly excavated and an embankment is constructed to retain the water. It is most common type of farm pond and generally constructed across the stream or nalla. A site which has already depression Y gentle to moderate slope, is usually best suitable for embankment type pond construction.
3. **Sankan pond :** It is dug cum embankment type pond constructed in the narrow valley portion of stream or nalla particularly at the site, which has already depression. These ponds are being widely constructed under NWDPRA for storing rainwater for the purpose of livestock's drinking & ground water recharge.
4. **Percolation tank :** Percolation ponds are small water storage structure constructed across natural stream and nalla to collect and impounding the surface runoff from the catchment during monsoon rains and store the water for longer time to facilitate infiltration & percolation of water into the pub surface to rise the groundwater level in the zone of which influence of pond.
5. **Off stream pond :** Off stream ponds are constructed by the site of ephemeral stream which flows seasonally. In this of ponds season rainwater is stored which needs a suitable arrangement for safe convenience of stream flow into the pond.

Excavated farm pond are of three types according to their shape:

i) Square ii) Rectangular iii) Circular

Among these, circular pond has higher storage capacity and has least circumferential length for a given area

i) Shaping of rock surface area and compacting the top layers of clays soil by which collect more runoff.

ii) Application of sodium salt to the soil, sodium compound disperse and swell the clay particles present in the soil.

iii) Surface treatment of soil with Bitumen & Asephate was found to be effective for storage of runoff water.

iv) 100% runoff harvesting efficiently could be achieved by covering the land surface by farm pond.

Material which can be used to reduce the percolation : These are the some variety of materials viz: natural clays, bitumen, soil cement mixture, stone or bricks in a cement materials, cement concrete, rubber and polythene sheets.

Evaporation retardants in farm ponds : Several anti-transparent like: plan residues, oil emulsions, fatty alcohol, gum mixtures, polythene, etc. could be used for prevention of evaporation losses.

Care maintenance of farm ponds : The farm ponds may be silted up fast, hence periodic digging to restore the original storage capacity is required. In order to know the depth of water collection in the farm pond and to also know the volume of water stored, can be used depth gauge and it should be installed in the pond.

Use of plastic mulch in pond lining : Low density polythene film (LDPF) should be used for lining of pond & it has found effective to reduce the seepage of water from pond and it works as a barrier between the soil and water for pond lining polythene film lining material should be 200 to 500 u thick.

6. **Spring or creek fed pond :** These are those where a spring or creek is source of water supply to the pond. These pond are generally constructed in hilly areas, where natural spring or creeks are available.

Estimate the volume of rainwater harvesting storage structure for watershed areas.

Establishment of volume of pond: The volume of excavation required, which can be estimated accurately by Prisonatial Formula:

$$V = \frac{A+4B+C}{6} \times D$$

Where,

V= Volume of excavation (m^3)

A= Area of excavation at ground surface (m^2)

B=Area of excavation at mid depth point (D/2) in m^2

C=Area of excavation at the bottom of pond (m^2)

(D)= Depth of pond (m)

Exercise: Estimate the volume of exaction required to construct a dug out pond with an average depth of 3.0 m, bottom width 9.0 m and bottom length 24.0m. The side slope adapted 2:1.

The volume of excavation required $V = \frac{A+4B+C}{6} \times D$

Solution:

Top Area (A):

Top length = 24 + (3 x 2) x 2 = 36.0 m

Top width = 9 + (3 x 2) x 2 = 21.0 m

Top Area = 36 x 21 = 756.0 m^2

Mid Area (B):

Mid length = 24+ (1.5 x 2) x2 = 30.0 m

Mid width = 9 + (1.5 x 2) x2 = 15.0 m

Mid Area: 4B = 4 x 30 x 15 = 1800m^2

Bottom Area (C):

C = 9.0 x 24.0 = 216.0 m^2

Total Area : A+4B+C = 756+1800+216 = 2772 m^2

$$V = \frac{2772}{6} \times 3 = 138m3$$

Answer : Volume of excavation : 1386 cubic metre.

□□□

Chapter 3

Doable Dryland Technologies for South-East Rajasthan

A.K. Kothari, J.K. Balyan and Devyani Dodia

Dryland Farming Research Station, (Maharana Pratap University of Agriculture and Technology, Udaipur) P.B.No. 62, Arjia, Bhilwara-311 001 (Rajasthan)

Introduction

Agriculture is the single largest livelihood sources in India with nearly two-thirds of people depend on it. Rainfed agriculture is as old as agriculture itself. Growing of crops entirely under rainfed conditions is known as dryland agriculture. About 70% of rural population lives in dry farming areas and their livelihood depend on success or failure of the crops. Dryland Agriculture plays a distinct role in Indian Agriculture occupying 60% of cultivated area and supports 40% of human population and 60 % livestock population. The contribution (production) of rainfed agriculture in India is about 42 per cent of the total food grain, 75 per cent of oilseeds, 90 per cent of pulses and about 70 per cent of cotton. By the end of the 20th century, the contribution of drylands will have to be 60 per cent, if India is to provide adequate food to 1000 million people. Hence, tremendous efforts both in the development and research fronts are essential to achieve this target. Dryland Farming Research station, Arjia, Bhilara is a constituent unit of Maharana Pratap University of Agriculture & Technology, Udaipur (Rajasthan). The research station was transferred to the University from State Government in the year 1976 and this farm was used for seed production. Research work on regional problem was initiated in 1981. Looking to the work of dryland agriculture All India Co-ordinated Research Project for Dryland Agriculture was started in 1976 at Udaipur and shifted to Arjia in 1984 along with an Operational Research Project for testing, refinement and dissemination of dry land technologies. This centre is one of the 25 AICRPDA centers spread over the rainfed regions of the country with it are headquarter at CRIDA, Hyderabad. Looking to the work of dryland agriculture name of this station was changed from Agricultural Research Sub-station to Dryland Farming Research Station in the year 1999 by the Hon'ble Vice-Chancellor, MPUAT, Udaipur.

Domain Characteristics

This centre is located in sub-humid southern plain and Aravalli hill zone (IV A) of Rajasthan. It covers geographical area of about 3.37 m ha which comprises of 9.84 per cent of total geographical area of the state. It spreads over Bhilwara, Rajsamand, Chittorgarh and parts of Udaipur and Sirohi districts of Rajashtan. Bhilwara district covers about 1.04 m ha area constituting about 31 per cent area of the zone. The mean annual rainfall of zone is 657.7 mm. The zone has 6.64 lac ha area under rainfed farming, which is 78% of the total cultivated area in the zone. The moisture deficit index (MDI) of Bhilwara district is (-) 53.1. The number of rainy days are 33. The frequency of drought in the zone is twice in 5 years. Inceptisols (50%), alfisols (36%) and entisols (5%) occupy a major area of the district. Soils of the district range from sandy loam to clay loam in texture. There are 15.26 lacs cultivators and 2.98 lacs agricultural labourers. Agriculture is the main occupation of the people in the zone. The percentage of workers population to the total population in the zone is 44.42 as against state average of 42.11 per cent. The average size of operational land holding is semi- medium and is 2.15 ha per cultivator. Cultivators constitute the maximum (52.12 per cent), working force, while the population involved in house hold industry is the minimum (2.58 per cent) where as agriculture labourer's share is 10.18 per cent only. Number of holding in the zone constitutes 5.84 per cent of the state. The total livestock population in the zone is 71.08 lakh, of which cattle and buffalo together contributes 48.15 per cent.

Maize is the principal crop and occupies more than 50% of the total cropped area in the zone. Sesame, sorghum, blackgram, greengram and groundnut are other important *kharif* crops. Wheat, mustard, barley and chickpea are principal *rabi* crops. The other emerging crops in the region are cotton, clusterbean, soybean and pearlmillet. Maize-fallow, maize- blackgram, maize- wheat, maize-groundnut, maize + blackgram mixed cropping and maize-mustard are the dominant cropping systems. Anole, orange, pomegranate, guava are the major fruit crops. Cumin and garlic are the important spice crops. Opium and ashwagandha are major medicinal plants. Nadi system of cultivation extends all over the zone. Animal husbandry is the main allied enterprise and cattle dominate the livestock population.

Major Research Achievements

1. Resource characterization

- Detailed analysis of long term rainfall data (1960-2016) was carried out to determine the variation in different parameters viz.; onset and withdrawal of south west monsoon, length of growing season, rainy days, frequency of dry spells, high rainfall events, crop water balance, etc.
- A criterion for Onset of monsoon has been developed and dynamic crop planning based on onset of monsoon has been recommended.

2. Rainwater Management

A. In-situ rainwater management

- Flat sowing of maze and ridging after first interculture was found as the most effective in-situ soil moisture conservation measure.

Ridging after sowing in maize

- On sloppy lands, growing of maize in lower half (receiver) and sorghum fodder in upper half (donor) area would help to sustain the maize yield during drought years.

- Use of vertical mulch at 10 m interval or dead furrow found efficient technique to enhance maize yield and improve the soil health under dryland ecosystem.
- Compartmental bunding of 5m X 5m has been found effective for in-situ rainwater conservation.
- Chiseling at 4 m interval once in five year has been found for breaking the plough pan and improving the infiltration rate.
- Taramira crop should be taken, if late rains sustained.

Chiseling to break plough pan

B. Ex-situ rainwater management

- The rainfall- runoff study revealed that 250 cum of runoff are available for rainwater harvesting and structure should design accordingly.
- Application of 5 ha cm harvested rain water as early supplementation followed by another 5 ha cm at 45 days later during post-monsoon period was found suitable for higher productivity of pigeonpea based intercropping systems.

Farm pond to harvest rainwater

- The harvested rainwater should be used for life saving irrigation in maize + black gram of groundnut +sesame intercropping system as compared to respective sole crops and more water is available, then it should be used for growing of mustard or vegetables like Kachari, coriander green leaves and brinjal.

3. Crops and cropping systems

- Intercropping systems, namely maize + blackgram (2:2), groundnut + sesame (6:2) , Maize / Groundnut + pigeonpea (1:1), sorghum + green gram (2:1), castor + green gram (1:2) during kharif and Intra cropping of gram + mustard (4 m apart across gram) during rabi were found most productive under dry land conditions.

Intercropping of Maize+Blackgram (2:2)

Intercropping of Groundnut+Sesamum (6:2)

- Under varying monsoon situations Cereal dominant system (CDS: 2 row Maize: 1 row Blackgram: 1 row Sesame) gave superior yield.
- Seed priming of maize with KH2PO4 @ 1% solution, enhanced the germination and produced higher grain yield of maize. Seed priming of maize reduces the germination period and found effective to boost up the initial growth of seedlings and finally the yield.

Cereal Dominant System

Detasselling in Maize

Seed priming

- Removal of tassels from 50% plant population between 3 to 7 days after emergence increased the yield of maize and also provides green fodder.
- Strip cropping of blackgram in 2/3 area and Maize chari (fodder) in 1/3 area was found good to meetout the food and fodder requirement as compared to sole maize.

4. Integrated nutrient management

- Application of 15 kg N through organic compost or green leaf and 10 kg N/ha through inorganic fertilizer was found to sustain the crop yield of maize and blackgram grown in strips after three cycles of maize- blackgram rotation.
- Application of 120 kg sulphur ha^{-1} through gypsum to groundnut + sesame intercropping to increase the productivity of oilseed crops.

- Zinc is the most limiting micro-nutrient in maize production system. The productivity of maize can be improved with the application of zinc sulphate @ 25 kg/ha along with recommended dose of N and P (50 kg N & 30 kg $P_2O_5ha^{-1}$).
- Spray of zinc sulphate @ 0.5% + NPK (19:19:19) soluble @ 2% at knee high and before tasselling stage was found effective to mitigate the dry spell stress in maize crop.
- Application of 40 Kg K_2O ha^{-1} and 30 Kg $MgSo_4.7H_2O$ ha^{-1} increased availability of K and Mg for the maize crop.
- Application of RDF of N and P with Azotobacter + PSB seed treatment and spraying of vermiwash @ 7.5 per cent at flower initiation and pod formation stage increased seed yield of Taramira crop sown on conserved moisture/with pre sowing irrigation.
- Application of 25 kg N through FYM and 25 kg N through Inorganic fertilizer + 30 kg P_2O_5 was found superior and increases the organic carbon, available N, available phosphorus, available Potssium and available soil micro nutrient viz.; iron, copper, manganese and zinc under long term experiment in maize - blackgram cropping system.

5. Organic Farming

- Application of vermicompost @ 1 t ha-1 + 0.35 t Rock phosphate (équivalent to 60 kg P ha-1) was found productive and profitable in groundnut crop.

Vermicompost
@ 1 t ha-1 + 0.35 t Rock phosphate
(équivalent to 60 kg P ha-1)

Poultry manure
@ 1.7 t/ha + BD 500 + BD 501

- Application of poultrymanure @ 1.7 t/ha along with biodynamic manure 500 + BD501 gave significantly higher seed yield of mustard and B: C ratio as compared to farmyard manure and vermicompost.

6. Evaluation of improved varieties

- Three horsegram cultivars viz.; AK-21, Pratap Kulthi-1(AK-42) and an extra early maturity cultivar Pratap Kulthi-2 (AK-53) had been developed at this research station and released in 1999, 2005 and 2015, respectively from Central Variety Release Committee for cultivation in dry land areas of Northern India.

Horsegram AK53 (Pratap Kulthi-2)

- Promising cultivars of major crops, namely Maize (PEHM-2, PM-3, PM-5), Sorghum (CSH-17, Pratap 1430 & CSV-15), Sesame (RT-351 and RT-125), Groundnut (TG 37A, GPBD-4 and PratapMungfali-2), Blackgram (PU-31, Pratap urd-1, KU-96-3, Vallabh U-1 and RBU- 38) Greengram (SML-668, RMG-492), Horsegram (AK-21, AK-53 and AK-42), Clusterbean (RGC- 936, RGC- 1031, RGC- 1038, & RGC- 1055), Soybean (JS-335 and JS-75-05), Gram (ICCV-10 and Dhohad yellow) and Dohad Yellow), Taramira (RTM- 314 and RTM- 2002) were identified for dry land areas of the zone.

- A community seed bank had been developed for Groundnut variety TG-37 A and community seed village for maize vairiety PM-3.

Groundnut variety TG-37A

7. Farm mechanization

- Arjia pora, Arjia wheel hoe, Arjia two row seed drill and Arjia intercropping seed drill had been developed at this research station and subsidized by the State Government.

Arjia Wheel Hoe

Arjia two row seed drill

- A Custom Hiring Center was established by the research centre in one of the NICRA village to improve the timeliness and precision in different farm operations with a participatory mode. The improved implements viz.; Reversible Disc plough , M B Plough, Rotavator, Rota-til-drill, Intercropping seed drill , Multi crop seed drill, Two row seed drill, Hand seed drill, Arjia wheel hoe, Single row power weeder (Four stroke), Single row power weeder (Two stroke), Three row power weeder, Battery operated power sprayer, and Chaff Cutter (electric). Farmers have appreciated the first time introduction of tractor drawn multi-crop seed drill for groundnut, single row power weeder for intercultural operation and battery operated power sprayer.

Further, they have also appreciated the improvement in availability of improved implements for increasing the efficiency and precision of different farm operations. A committee of farmers has been constituted to operate the same. It has been used 584 farmers in an area of 196.0ha.

Arjia intercropping seed drill

Table : Details of implements used under customer hiring centre

Type of machinery/ equipments used	Number of equipments	Type of farm operation	Number of farmers/	Area covered, ha	Diesel consumption (lit/hr/operation)
Reversible Disc plough	One	Primary tillage	20	8.0	3.2
M B plough	One	Primary tillage	40	16.0	4.2
Intercropping	One	Sowing	80	32.0	1.3
Two row seed drill	three	Sowing	8	8.0	1.25 ha / day/bullock power
Arjia wheel hoe	Ten	Interculture	216	56.0	0.09 ha/hr/person
Single row power weeder (Four stroke)	One	Interculture	56	40.0	0.5
Single row power weeder (Two stroke)	One	Interculture	56	32.0	0.65
Battery operated power sprayer	Three	Spraying	248	60.0	8 hrs per charging
		Total	584	196.0	

Rota-till-drill

Rotavator

8. Alternate land use system

- Outer ring of the Aonla basin should be used to grow pumkin at six places to have an additional return.This space can also be used for growing of sunhemp for green mannuring.

Growing of cucurbits in Aonla basin to get additional returns

Aonla Plantation

- Sowing of *Cenchrus* grass between the inter row space of exiting *neem* tree by Mud ball/pallet method with 20 kg N+30 kgP_2O_5/ ha, increased dry grass yield.

Improve productivity of Pasture land through Cenchrus grass

- Medicinal crops like Ashwagandha (Withania somnifera) and Senna (Cassia angustifolia) can be successfully intercropping interspaces of aonla. Among vegetables, pumpkin was found most profitable vegetable in interspaces of aonla. During Rabi season, taramira is the most profitable crop to utilize conserved soil moisture in aonla orchard.

- Bio-diverse farming system model suitable for dry land farming was evaluated and demonstrated to the farmers for improving and sustaining the productivity of dry land.
- Interspaces between ber plants can be economically utilized by growing Cenchrus setigerus grass.

Efficient use of interspace by growing of pulresin Aonla

- Trench planting across the slope with 5 m runoff length proved most effective micro site improvement technique for ber.

Trench planting of Ber

- Acacia tortilis was the most suitable tree species for silvi-pastoral system on marginal lands.

Conclusion

The challenges of the present millennium would be to produce more from drylands while ensuring conservation of existing resources. Hence, new strategies would have to be evolved which would make the fragile dryland ecosystems more productive as well as sustainable. In order to achieve evergreen revolution, we shall have to make grey areas (drylands) as green through latest technological innovations. By adopting a holistic approach for management of rainwater, crop productivity in rainfed areas can be increased. Strengthening of research infrastructure on rainwater management including in-situ conservation and ex-situ harvesting of water in this area are of paramount importance.

❑❑❑

Chapter 4

Four Water Concept and its Application in Increasing Productivity of Rainfed Agro Ecosystem

R.C. Purohit

Director, JNU, Jaipur, Rajasthan

Introduction

The Four Waters Concept technology was first adopted in Nampi Project in Hebei province in China and later in the new watersheds in Andhra Pradesh between 2000-2004. According to Sh T. Hanumantha Rao, the former engineer-in-chief and former consultant to UN "Through the Four Waters Concept, groundwater can irrigate three crops at a meagre cost of Rs. 15,000 per hectare, whereas a new major irrigation project in Telangana costs Rs 8 lakh per hectare and provide water for only one irrigated dry crop during the rainy period of the Kharif season. Due to this advantage, no new major or medium irrigation project is required to be executed in the country, he claimed. This technology emphasizes on water spreading techniques to facilitate recharge to ground water, in the upper areas of watershed, through mini percolation tanks instead of the conventional check dams in main valley.

Four water concept in Rajasthan region

Rajasthan is the largest state of the country has an area of 343 lac hectare. Out of this only 168 lac hectares is arable land. State has 101 lac hectare of land under wasteland. The Aravali range extends from South-West to North-East divides the state in two parts; on its best there is Thar desert which has an area around 60% of the state. The annual rainfall is 100 mm in dry hot West to 900 mm in South-East.

Generally, out of five years, 3 years are drought affected means untimely, uncertain & varied distribution of rainfall, this is responsible for the unstable crop production. Large part of the rainfall goes waste due to high intensity of rainfall in small time. Proper utilization of runoff does not take place due to lack of Water Harvesting Structures in the watershed area results continuous decline in water level of the wells. Normally in Rajasthan, duration of rainfall is large which has adverse effect on crop production. Due to decline in crop production of farmer and conversion of arable land to wasteland, there is scarcity of fuel, fodder, milk resulting for their weak socio- economic condition. Scarcity of water is the main reason for this situation. Watershed Department, if work for preparation of an Action Plan in co-ordination with other departments, it will definitely be able to provide water for drinking purpose and for assured irrigation. As State has 75 percent rural population which depends on agriculture and livestock for their livelihood, hence in light of these conditions of state, there is need of Mukhya Mantri Jal Swawlamban Abhiyan.

Four Waters concept can come very handy in the drought-prone Rajasthan for countering drought and subsequent damage. This low-cost technology can increase recharge of groundwater by four times and provide thrice the benefit than conventional models. The methodology of the Four Water concept revolves around the harvesting of available runoff (rainwater, groundwater, underground water & in situ soil moisture) in rural areas by treatment of catchment, proper utilization of available water harvesting structures, renovation of the non-functional water harvesting structures & creation of new water harvesting structures.

Application of Four water concept in increasing productivity of rainfed agro ecosystem in India: The campaign, inaugurated by Mrs. Raje at Jhalawar, and at other places across the Rajasthan State by the Ministers, aims at benefitting 21,000 villages by 2020, with 3,000 villages benefitting this year and 6,000 villages per year for the next three years.

Objectives of MJSA

1. To ensure effective implementation of water conservation and water harvesting related activities in rural areas through convergence of schemes of various departments keeping in view the existing guidelines.

2. Implementation of works through people's participation by motivating villagers & beneficiaries.
3. To ensure effective implementation of water conservation and water harvesting related activities in rural areas through convergence of schemes of various departments keeping in view the existing guidelines.
4. Implementation of works through people's participation by motivating villagers & beneficiaries.
5. At village level in Gram sabha after determination of usage of water like drinking, irrigation, livestock & for other commercial purpose & preparing water budget keeping in view of water available from various resources, accordingly works will be identified and proposal get approved for the preparation of action plan of mission.
6. At village level in Gram sabha after determination of usage of water like drinking, irrigation, livestock & for other commercial purpose & preparing water budget keeping in view of water available from various resources, accordingly works will be identified and proposal get approved for the preparation of action plan of mission.
7. Harvesting of available runoff (rain water, ground water, underground water & in situ soil moisture) in rural area by treatment of catchment, proper utilization of available water harvesting structures, renovation of the non-functional water harvesting structures & creation of new water harvesting structures.
8. Development of Forest, land, water & fauna keeping watershed/cluster/index as a unit for natural resource management.
9. Permanent Solution of drinking water by making the village self-sufficient in terms of water.
10. To increase the Irrigated area through water harvesting & conservation.

Application of four water concept in increasing productivity of rainfed agro ecosystem around the world : The 'four water concept' has been tested and refined through several years in the Nampi experimental station in Hebei province in China. A pilot area of 23,600 ha watershed was selected for this purpose. Yields of crops increased by 77%. Through the average annual rainfall is 550 mm, it was possible to irrigate one light irrigated crop in dry season to an extent of 43% cropped land. Experience gained in Xiong Xian county, in the central part of Hebie Province were similar. The water table which was previously at a dangerously low level was raised and kept between 2 m below ground level and 6 m below ground level. This result is substantially energy saving and increased yields.

Institutional Arrangements for applying four water concept

- State Rural Water Conservation Mission
- State level Directional Committee
- State level task group
- District level Monitoring Committee
- District Level Committee
- Block Level Committee
- Village Level Committee to prepare Action Plan

Conclusions

The 4 water concept provides a scientific and practical approach to develop a watershed to derive the maximum benefit from the land and water resources in the area. It arrests the trends of depleting the water tables year after year and would increase the water levels appreciably over a period of time. Loss of soil moisture would be prevented and rainfed crops would give better yields. Bio- fertilization out of mulching would improve the structure and fertility of soils. So, it supports agro ecological systems.

Source

It is possible to manage more efficiently the water available in the 'four water concept', namely rainwater, soil water, groundwater and surface water. The advantages of this concept are striking and hence, it is suggested that the same is adopted for all watershed development programs in India.

References

WebNotefrom:https://books.google.co.in/books?id=4C4y00 LwQecC&pg=PA70&dq=watershed+and++four+water+concept+ and+its+application&hl=en&sa=X&ved=0ahUKEwiE3JeLwqfYAhXKv48KHWnED_kQ6AEIKDAA#v=onepage&q=watershed%20and%20%20four%20water%20concept%20and%20its%20application&f=false

□□□

Chapter 5

Integrated Watershed Management

M.S.Shirahatti
Chief Scientist, Regional Agricultural Research Station, Vijayapura

India has made an appreciable progress in creating irrigation potential. However, it is still insufficient to meet the long term requirement of irrigation. The ever increasing population has put tremendous pressure on food demand. Every unit of available land resource and other critical inputs needs to be exploited to reap maximum benefits. In feature the most critical input happens to be water. For efficient utilization of precious natural resources like soil and water, effective soil and water conservation measures is the need of the hour. Soil and water conservation measures can be broadly grouped into two categories, i.e., agronomical practices and engineering measures. In soil and water conservation programmes, the agronomical practices are counted as second line of defense, the first being mechanical or engineering measures which are employed to arrest the soil erosion immediately. The role of agronomical measures is more economical long lasting and effective. Always it is advisable to use agronomical practices, but when its use is either inadequate or not sufficient to achieve the goal of soil and water conservation, and then use of mechanical measures is recommended.

Soil and water conservation measures are grouped as

a) Agronomical and biological measures

i) Water erosion control measures includes contour farming, strip cropping (border strip and contour strip cropping), conservation cropping system, mulching and crop residue management, critical area planting, vegetative barriers or live bunds, land smoothening, ridges and furrows/dead furrows.

ii) Wind erosion control measures, such as vegetative cover, strip cropping, stubble mulching, tillage practices, sand dune stabilization, windbreaks and shelter belts

b) Mechanical or engineering measures

The mechanical or engineering measures can be broadly divided as bunds and terraces.

Terrace level practices

I **Bunds :** Bunds are used for retaining the water by creating obstruction and thus, to control erosion. Bunds are simply embankment like structures constructed across the land slope. When they are constructed on the contour of the area called contour bunds and when a grade is provided to them, and then they are known as graded bunds.

i) **Contour bunds :** The formation of bunds passing through the points of equal elevation is called contour bunds. The contour bunds can be constructed on all types of permeable soils, such as alluvial, red, laterite soil, shallow and medium black soils but clayey or deep black cotton soils are strictly avoided for these structures as these soils have the problem of crack development, which causes the bund failure against its function. In addition clayey, soils are also borne the problem of water logging condition near the bund section, which makes the bund construction unfeasible.

Main functions of contour bund

1. Reduces the length of slope, which in turn to reduce the soil and water erosion.
2. Helps to impound the water at upstream portion and permits more water to recharge into the soil profile.

The limitations of contour bunds

1. This practice is suitable for those areas which receive the annual rainfall upto 500 mm and existing soil involves greater permeability.
2. Bunds cannot be adopted, when land slope is greater than 6 per cent.

2. Zing conservation bench terraces

These are trapezoidal earthen embankments constructed on contour with contributing/ donor area and leveled/ receiving area with or without surplussing arrangement depending on soil type, rainfall and desired area of water submergence.

Specifications

- Bund cross section - 1.0 to 1.2 sq.m.
- Vertical interval - 1.0 to 1.5m (depending on land slope)
- Level to unlevel proportion - 1:3 to 1:4
- **Area of applicability :** Can be successfully practiced in medium to deep black soils having infiltration rates more than 8mm/hr. on lands having slope upto 5% in areas receiving low to medium rainfall.

3. Contour border strips

These are level strips with no slope across the width constructed parallel to a contour key line with suitable drop structure in the water way at the end of each strip. Two contour border strips are separated by a low border ridge.

Specifications

- Vertical interval between two ridges - 0.3m
- Border ridge section - 0.24sq.m. (to stabilized by planting grasses)
- Length of border strip - 100 to 150 m.
- **Area of applicability :** Can be successfully practiced in medium to deep black soils having infiltration rates more than 8mm/hr, on lands having slope upto 3% and in areas receiving low to medium rainfall (around 750mm).

4. Graded bunds with strip leveling

These are trapezoidal earthen embankments constructed on a grade across the major slope so as to intercept runoff and lead the excess water through a leveled strip constructed on 0.2 to 0.4% grade on the upstream side of the bund.

Specifications

- Bund cross section - 1.5 to 0.6 sq.m.
- Vertical interval - 0.75 to 1.0m
- Grade - 0.2 to 0.4 per cent
- Level strip width - 10 to 15m
- Protected water way is a pre-requisite.
- **Area of applicability :** Can be successfully practiced in areas receiving annual rainfall of more than 750 mm. where runoff is high and surplussing of excess runoff is essential. They are also fit for adoption on clayey black soils receiving less than 450 mm of annual rainfall and having infiltration rates less than 8 mm/hr.

5. Graded border strips

These are the strips with no slope across the width constructed on grade, parallel to a graded key line with suitable drop structures in the waterway at the end of each strip. Two graded border strips are separated by a low border ridge.

Specifications

- Vertical interval between two ridges - 0.3m
- Border ridge section - 0.24 sq.m. (to be stabilized by planting grasses
- Border grade - 0.2 to 0.3%
- Length of border strip - 100 to 200m.
- **Area of applicability :** Can be successfully practiced in medium to deep black soils having infiltration rate less than 8 mm/hr and in areas receiving rainfall of around 750 mm on lands having slope up to 3%.
- **Terraces :** Formation of terraces involves the construction of embankment or ridge and steps like structure across the land slope to check the flow of surface runoff and to reduce the soil loss. In this system, the effective length of land slope is reduced to a large extent. Terracing is not feasible in flat land.

The main features of terraces are:

a) These are constructed across slope to intercept the surface runoff and convey it to a suitable outlet at non-erosive velocity.

b) They reduce the length of slope by splitting the slope length in different parts.

c) This practice is adopted for soil and water conservation in that area where land slope is greater than 10 per cent.

Types of terraces : These are classified into three main classes, namely diversion, retention and bench terraces:

i) **Diversion terrace :** It is used for intercepting the overland flow from hilly slopes and channel it across the slope to a suitable outlet i.e. grassed water way.

ii) **Retention terrace :** These are level terraces used particularly when water is required to conserve by making storage on hill sides.

iii) **Bench terrace :** These are constructed in form of alternate series of shelves and risers, used to cultivate the steep slopes.

These practices are known to conserve the rainwater to the extent of 60-70% and reduce the soil erosion by 65-75% and increase the crop yield by 35-40%.

II Water harvesting structures

Farm ponds

Farm ponds are nothing new to Indian agriculture and these are the bodies of water, made either by constructing a dam or an embankment across a water course are by excavating a pit or dug out. The dug out farm ponds are a common feature of relatively flat terrains. In arid areas where the water table is usually very deep, the surface runoff is the source for feeding the ponds. Farm ponds are potential source of supplemental irrigation and drinking water in rainfed areas. Before taking a decision on making a farm pond, it is necessary that the detailed features of the contributing area, suitable site, possible utilization of the stored water and economics in terms of benefits are to be studied.

Check dams

The main objective of construction of check dams is to stabilize the gully and bring back the eroded land for cultivation in addition to storage of harvested water. Check dams are constructed across small streams having gentle slope. The site selected should have sufficient thickness of permeable bed or weathered formation to facilitate recharge of stored water within a short period of time

Percolation tanks / Nala bunding

These are the structures constructed in waste lands along the nala to store the harvested water.

Mini barrages and reservoirs

These are the structures constructed in the rivers to store the harvested water.

Sand Dam

Water harvesting by use of sand dam consists of constructing a dam across the valley or depression for the purpose of reducing the flow velocity of water mixed with sand. By this, silt load of runoff tends to deposit over the bed. After some time, the bottom of valley is raised due to deposition of sand. Silted part of the area known as sand reservoir. Water flowering through the valley, fills into the pore of the area known as sand reservoir. Although, the storage capacity of this reservoir is small due to the space covered by the sand, but this is partly compensated by the water stored in the pore spaces, which are larger in size.

Collector pipe Network

Subsurface Dam

It consists of subsurface vertical which are constructed across the valley, down the bed. These barriers intercept the flowering water within the alluvium. The water flow through the valley is stored in sub-surface reservoirs, created by barriers. In this dam, the evaporation loss is minimum. The subsurface dams or barriers are generally constructed with clay, stone, masonary, concrete or steel materials, etc. the water stored in these dams are utilized by gravity or shallow wells or by bore holes.

Small Earthen Embankments

These are constructed by forming dam or embankment across the valley or depression of the catchment area. Storage capacity of the reservoir is determined on the basis of water demands and available surface runoff from the catchment. In the condition, when heavy usage of water is expected, then the It's storage capacity of reservoir must be kept sufficient so that it can fulfill the demand for more than one years.

Recharge shaft

RCC tank

Dug outs

A dug out is a pit excavated to provide storage below the ground surface. These are usually located in wide flat drainage channels at the low point or at any other place on fairly flat topography, where water can be collected and impounded below the ground surface. A series of dug - outs help in the recharge of ground water as well water is made available for irrigation of neighbouring lands.

Roof top water harvesting and artificial recharge to ground water

Rain water harvesting is the technique of collection and storage of rainwater at surface or in sub-surface aquifer, before it is lost as surface runoff. The augmented resource can be harvested at the time of need. Artificial recharge to ground water is a process by which the ground water reservoir is augmented at a rate exceeding that under natural conditions of replenishment. In urban areas, rain water available from roof tops of building, paved and unpaved areas goes waste. This water can be recharged to aquifer and can be utilized gainfully at the time of need.

References

1. Ali, Shakir, Sethy, B.K., Singh, R. K, Parandiyal, A.K., and Kumar, A. (2017). Quantification of hydrologic response of staggered contour trenching for horti-pastoral land use system in small ravine watersheds: a paired watershed approach. Land Degradation and Development, 28(4): 1237-1252. DOI: 10.1002/ldr.2517.
2. Prathapar, S., Dhar, S., Rao, G.T. and Maheshwari, B. (2015). Performance and impacts of managed aquifer recharge interventions for agricultural water security: A framework for evaluation. Agricultural Water Management, 159: 165–175.

❑❑❑

Chapter 6

Moisture Conservation Practices for Rabi Sowing

V.S. Surakod
AICRP for Dryland Agriculture, RARS, Vijayapura

Land and water are the most vital natural resources of the country and these are under tremendous stress due to ever increasing biotic pressure.The optional management of these resources with minimal adverse environment impact is essential not only for sustainable development but also for human survival.Watershed is an ideal unit calling for ensuring continuous benefit on sustainable basis. The concept of watershed management is essentially adoption of soil and water conservation practices in the watershed.The aims of these practices are proper landuse, protecting land against all forms of deterioration, building and maintaining soil fertility, conserving water for farm use, proper management of local water for drainage, flood protection and sediment reduction and increasing productivity from all landuses.Watershed management must consider the social, economic and institutional factors operating within and outside the watershed area.

Some of the improved moisture conservation practices for watershed management are mentioned below:

1. **Fall ploughing :** It comprises ploughing the land soon after the harvest of kharif/rabi crops or immediately after the first showers mainly to make the soil surface smooth in a receptive condition which facilitate for entry of water into the soil by reducing the run-off and conserve the water. Apart from this, it is also known to control the weeds and pests, and improves the soil physical properties.
2. **Primary tillage operation (Deep ploughing) :** It consists of operating a tractor MB plough or a bullock drawn country plough

for opening up or loosening of soil. It helps to remove the abnoxious weeds and provided better physical conditions for holding water in soil profile and easy penetration of roots of plants. This can be done on all types of soils under varying rainfall conditions. Deep ploughing is done once in two–three years during summer in medium to deep black soils of Karnataka mainly to break the hard pan in soil profile. The clods formed due to this operation acts as barrier for the flow of runoff water and provide more opportunity time for rainwater to infiltrate into the soil. The crop yield increased to an extent of 50-60 per cent.

3. **Contour cultivation :** It is the type of cultivation practice carried out along the contour key lines mainly to build up temporary water storage capacity in furrows across the slope. Based on research at Bijapur, contour cultivation could increase the sorghum grain yield by 28.6% compared to conventional practice.
4. **Compartment bunding :** Compartment bunds are formed with the help of bund former during the month of July. Small bunds of about 10-15 cms height will be formed in both directions and these small bunds act as a barrier for run-off water and provides more opportunity time to run-off water in each compartments to infiltrate into the soil profile. Thereby, it increases the soil moisture and useful to grow the crops on receding soil moisture. This practice is more benefitted in medium to deep black soil where the infiltration is very low. It is also known to control soil erosion more efficiently. The size of compartment bunds depends on the slope of land and type of soils. In large scale study conducted at Kumatagi watershed of Bijapur district, the grain yield of rabi sorghum was increased by 37.6%.

Deep ploughing

Con tour cultivation

Contour cultivation

5. **Tied ridges :** Ridges and furrows are formed during the month of July and they are tied at regular intervals. The spacing between furrows are fallowed based on spacing used for the crops. They act as a barrier and a store structures for runoff water and thus, the water get infiltrate into the soil due to retention of water for longer time in each furrow. Thereby, it increases the soil moisture and useful to grow the crops on receding soil moisture. This practice is more benefitted in medium to deep black soil where the infiltration is very low. It is also known to control soil erosion more efficiently. The grain yield of rabi sorghum was increased by 32%.
6. **Scooping :** It is the practice of opening of small pits at definite intervals on the surface of land across the slope or along the contours. They are formed during the second fortnight of July. These small pits act as small storage structures for runoff water and thus reducing the soil erosion and increase the soil moisture content. This practice can be used more predominantly in medium to deep black soils where infiltration capacity is very less. It is very important to revive the scoops after each rain to facilitate better infiltration.
7. **Broad bed and furrows :** These seedbeds are made into wide beds of 1.05 m alternated with furrows of width 0.45 m. These are called broad bed and furrows. This method is quite suitable for black soils, when combined with dry sowing. The runoff from broad bed is collected in the furrows helps for retention of water in the furrows which increases the moisture content of the soil. The crops will be grown on the broad beds furrows will help to conserve the water.

Graded Border Strips

Graded Border Strips

Graded Bund

Zingg. conservation bench terrace

8. **Gravel sand mulching:** It is the practice of using gravel sand as mulch on the soil surface mainly to conserve more rain water in situ and avoid loss of soil moisture through evaporation. By adopting this practice, it is possible to take up double crops in the field. It is more significant practice particularly during the less rainfall year. Before application of sand the land is thoroughly ploughed and applied heavy amount of FYM to enrich and soil fertility. Nearly 275-300 tractor loads of sand are used to apply on land to reach 10 to 15 cm depth. The yield of greengram, rabi sorghum and chickpea increased to the extent of 200-250 percent. This practice is more popularly followed by medium to big farmers of Yerehanchinal village of Koppal district of Karnataka in Kurl soils.

Gravel sand mulc hing

9. **Pebble mulch :** It is the practice of retaining naturally available pebbles on the soil surface. Retention of pebbles acts as barrier for the flow of runoff water and provides more opportunity time for the rain water to infiltrate into the soil profile. Apart from this, it reduces evaporation. This practice increases the yield of greengram, rabi sorghum and sunflower to the extent of 85-200 per cent.

Ridges and Furrows

Tied ridges

Scooping

Compartment bunding

10. **Retention of crop residues and incorporation :** The residues left after the harvest of the crop have to be retained as it is on the soil surface. These acts as barrier and slows down the velocity of run-off water. These reduce the soil erosion and increase the soil moisture in the profile. The retained residues are incorporated into the soil by rotovator. The research evidence at Regional Agricultural Research Station, Bijapur proved that retention of sorghum stubbles/sunflower stalks on the soil surface increased the yield and net income by 20% and 30% over control.

11. **Incorporation of Green manure crops :** Some of green manuring crops like sunhemp, Glyricidia, greengram after the harvest of pods are incorporated into the soil with the help of rotovator. Incorporation of green leafy manures improves the soil, moisture holding capacity of the soil and also adds nutrients to the soil. In addition it also increases the physical, chemical and biological properties of the soil. The research evidence at Regional Agricultural Research Station, Bijapur indicated that growing of greengram during kharif and incorporation of residues produced less runoff by 79% and soil loss by 31% resulted in higher yield to the tune of 64% over traditional practice.

12. **Cover crops :** Cover crops are those crops which have fast initial growth and cover the land immediately which prevent the soil from heating action from rain and control the runoff and soil erosion. Vegetation is crucial in preventing productive soil and conserving rain water for sustaining life, soil and water needs to be preserved for crop production. Crops like greengram, black gram, horse gram and cowpea are some of the cover crop apart from yielding they also act as a cover crop.

13. **Set row cultivation :** Set row cultivation is the type of cultivation more popularly followed by the farmers of Rajkot and Surendranagar districts of Gujarat in medium to deep black soils for groundnut and cotton crops. Before opening of set rows, deep ploughing is done to break the hard pan in the soil profile. Then land is harrowed to bring the land to fine tilth. Set furrows are opened according to spacing followed for the crops. The well decomposed FYM @ 5-6 t/ha, green manure crops viz. glyricidia and crop residues are filled in the set furrows. Further the crops are sown in the set rows after receipt of rain. This type of cultivation helps to increase the efficient use of nutrients and moisture and reduce the cost of cultivation and increases the biological properties of soil.

14. **Wider row spacing and frequent deep intercultivation :** Adoption of wider row spacing ensures larger cross section of effective rootzone, which provides moisture for longer period. Frequent deep intercultivation reduces evaporation considerably by creating a dust mulch. This wider row also helps for better light interception, crops make use of the space between the rows. It also helps for repeated interculturing operations. It is very much suited to the crops which are more exhaustive for nutrients and moisture.

References

1. FAO (Food and Agriculture Organization) (1993). Land and Water Bulletin No. 1 Proc. Informal Workshop on Land and Water Integration and River Basin Mgmt., FAO, Rome.

2. Khan, Q.I., Venkataratnam, L., Rao, B.R.M., Rao, D.P., and Subrahmanyam, C. (2001). National classification and codification of watershed and river basin. Journal of Water Resources Planning and Management, 127(5). 306-315.

3. Prathapar, S., Dhar, S., Rao, G.T. and Maheshwari, B. (2015). Performance and impacts of managed aquifer recharge interventions for agricultural water security: A framework for evaluation. Agricultural Water Management, 159: 165–175.

❑❑❑

Chapter 7

Negotiating Drought through Engineering Interventions in Arid, Semi-arid and Sub-humid Regions

Virendra Kumar Srivastava

Former Professor and Head Deptt. of Soil and Water Engineering, Maharana Pratap University of Agriculture & Technology, Udaipur

Introduction

Several parts of the country usually witness serious drought one year or the other. In fact for the past few years the unprecedented drought has taken its toll on agricultural production. Some years it is certainly acute thereby resulting in tremendous shortage of drinking water particularly in the states of Gujarat, Rajasthan, West Madhya Pradesh, Odisha and Andhra Pradesh including Delhi and Haryana. Most of the dams and reservoirs in the region dry up. Given the fact that India is one of the most well-endowed nations in the world in terms of average annual rainfall, there is no reason why it should suffer from drought this year or any other year. Drought relief works, emergency water supply through tankers digging deeper bore wells are no long-term solutions. The most important lesson that we should learn from such crisis is to search out some long-term technological options to negotiate the drought and completely banish it in the years to come. The best option, however, is to trap water from monsoon rain and harness it, when it is needed most.

Potential of Rainwater Harvesting

Recognizing the fact that almost all the rain comes down in a few hours, our ancestors had learnt to harvest water in many ways:

1. They had harvested the rain drop directly from rooftops; they had collected water and stored it in tankas built in their courtyards.

From open community lands, they collected the rain and stored it in artificial wells called **kundis**.

2. They harvested monsoon runoff by capturing water from swollen streams during the monsoon season and stored it in **zings** in Laddakh, **ahars** in Bihar, **johads** in Rajasthan and **eris** in Tamilnadu, to name a few.
3. They harvested water from flooded rivers in places like north Bihar and West Bengal.

India's total land area is over 300 million ha. Let us assume that India's about 6 lakhs villages can harvest the runoff from 200 million ha of land, excluding inaccessible forest areas, high mountains and uninhabited terrains, that still gives every village on average access to 340 ha or a rainfall endowment of 3.75 billion litres of water. These calculations show the potential of rainwater harvesting is enormous and undeniable. There is just no reason whatsoever for thirst in India. Therefore, it is possible to drought proof the entire country. Not just drinking water, most of India's agricultural fields should also be able to get some irrigation water to grow less water-intensive crops every year through rainwater harvesting.

Community based rainwater harvesting

Let us look at the relevance of community or village-based rainwater harvesting from yet another point of view. The key component of water management is 'storage' especially in a country like India, where the monsoon gives us on average about one hundred hours of rain and then nothing for the remaining 8,660 hours in a year.

This water can be captured in:

a) Large reservoirs with large catchments by building large dams,

b) In small tanks and ponds with small catchments, or

c) By storing it in a way that it percolates down into the ground and gets stored as groundwater.

In fact, there is strong scientific evidence to show that village-scale rainwater harvesting will yield much more water than bigger or medium dams, making the later an extremely cost-ineffective and unscientific way of providing key water needs especially in dry areas.

Some very instructive lessons can be learnt from the work carried out by an Israeli scientist, Michael Evenari, who has produced the best corpus of knowledge on this subject from the bone-dry Negev desert where the average annual rainfall is a mere 105 mm.

At least the first two years of any water harvesting programme will have to be spent on social mobilization. Social mobilization is essential for the success of water harvesting for several reasons. Firstly, the community must be closely involved in the construction of the water harvesting structures to ensure that they are built with technical competence; that is, the site is chosen properly, the technical parameters are correct, etc.

Water Harvesting Technology Options

Water harvesting had been practiced from the primordial days in various ways. It is based on the following three basic criteria; source of water available, required storage duration and intended use of harvested water. On the basis of these criteria, the water harvesting techniques are classified as:

- Surface runoff harvesting.
- Underground or subsurface harvesting

Surface Runoff Harvesting

Surface runoff harvesting techniques are mainly classified as short term runoff harvesting and long-term runoff harvesting technique. Short term runoff harvesting techniques are designed to collect the 24 hr rainfall excess at 10 years recurrence interval. Field bunds, contour bunds, V-ditches, contour trenches, staggered trenches, etc. are the structures/measures which are commonly known as short term runoff harvesting techniques. These techniques are the most common and are widely adopted in several countries.

Long-term runoff harvesting is mainly done for having a water stock for livestock, drinking, irrigation, fish farming and ground water recharge. This is accomplished by constructing a reservoir or pond in the area:

- Catchment area should contribute sufficient amount of runoff.
- There should be some suitable collection site, where maximum water can be safely stored.

- There should be suitable locations, where the stored water could be saved by minimizing the various water losses, such as seepage and evaporation.

The most common type of long-term surface runoff harvesting structures are Farm ponds, Percolation ponds, Sunken ponds, Johads, Nadis, Khadins & Anicuts which are widely adopted in the country.

Farm ponds are small tanks or reservoirs constructed for the purpose of storing water essentially from surface runoff. Farm ponds are useful for providing supplemental irrigation to the crops, supply of water to livestock, fish production, etc. A large number of ponds constructed in the catchment area reduce the flood flows downstream. Farm ponds are also used for storing rainwater which is subsequently used for supplemental or life saving irrigation particularly in the condition of dry spells. Thus, a farm pond has a significant role in the area of rainfed agriculture.

Dugout Farm Pond

Dugout farm ponds are excavated at the site and the soil obtained by excavation is used to form an embankment around the pond. These ponds are generally constructed in the area having flat topography and are advantageous in the area where evaporation losses are high because it is constructed to expose a minimum water surface area in proportion to volume. These farm ponds are generally used where only a small supply of water is required. The size of the pond depends on the catchment area, soil type and expected runoff.

Embankment Type Farm Ponds

This type of pond consists of an earthen embankment or earthen dam, which is partly excavated, and an embankment is constructed to retain the water. It is most common type of farm ponds and generally constructed across the stream or nalla. A site, which has already depression and gentle to moderate slope, is usually best suited for embankment type pond construction.

Sunken Ponds : Are constructed in the narrow valley portion of stream or nalla particularly at the site, which has already depressions for the purpose of livestock drinking and ground water recharge.

Percolation ponds : Are constructed across natural streams and nalla to collect and impound the surface runoff from the catchments during monsoon rains and store it for longer time to facilitate infiltration and percolation of water into the substrata to raise the ground water level in the zone of influence of pond. Experiments have shown that the influence of percolation pond is 1.5 km and average rise of water table as 2.5 m (Sinha, 1997).

Spring or Creek Fed Ponds

Spring or creek fed ponds are those where a spring or creek is the source of water supply to the pond. These ponds are generally found in hilly areas, where natural spring or creeks are available.

Off-Stream Ponds

Off-stream ponds are constructed by the sides of ephemeral streams, which flows seasonally, to store seasonal water with a suitable arrangement for safe conveyance of stream flow into the pond.

Khadin

Khadins are structures on low lying lands where crops are raised by conserving rainwater received from the relatively impervious uplands with steeper slopes. Khadin bunds are constructed across the slope of the land in arid region, where amount of rainfall is very less and distribution is erratic. The ratio of Khadin submergence area to catchment area, depending on type of catchment is 1:12 to 1.15. The water collected is allowed to percolate, after which an assured post rainy season crop is grown.

Nadi

In arid regions, construction of nadi is an age-old practice of water harvesting. These are small excavated or embanked village ponds constructed for harnessing the meager precipitation to mitigate the scarcity of drinking water in the desert. These nadis hold water from two months to a full year after rains depending upon the catchment characteristics, amount of rainfall received, its intensity and distribution.

Tanka

Tanka is the most prevailing rainwater harvesting structure in Indian desert. The name "Tanka" is a local term for the underground cistern. The Tanka's are underground structure constructed to harvest surface runoff of the field as well as roof water for drinking purposes.

The traditional tankas are made by digging a hole of 3.0 to 4.25 m diameter in the ground and plastering it with lime mortar to a thickness of about 6 mm followed by a cement plaster of 3 mm thick. The top is covered with Ber thorns. The useful life of such structure is about 3 years. Traditional tankas are temporary and are subjected to leakage. The catchment area is not in accordance with the amount of rainfall required and runoff generated. The thorn cover does not prevent the water pollution and evaporation losses. Capacity of such tankas is also not sufficient to fulfill the demands of a family for water throughout the year. The improved tankas are now being constructed with cement masonry and its top is covered with long stone slabs, or RCC in which, a hole is kept for lifting water. These structures are lined and not at all subjected to leakage.

Anicut

Anicut is a weir structure constructed across the natural drain or nalla for impounding the water. It is a masonry structure, which intercepts the water from local catchment and stored it for optimum utilization. The word "Anicut" is a local name of this water harvesting structure commonly known in Rajasthan.

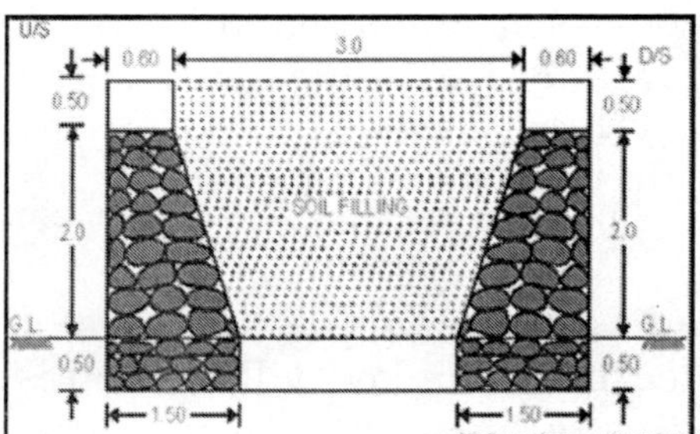

Pond - dry stone masonry

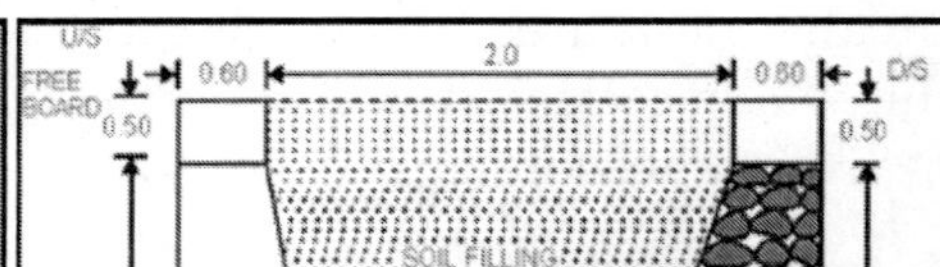

Pond-cement masonry

The water stored behind the structure can be used for supplemental irrigation to the crop as well as provide drinking water for human and livestock. This will also help in recharging of the wells located downstream side of the structure. Selection of suitable site of "Anicut" is very important as the cost of construction as well as storage volume depend upon the site.

Indigenous Water Harvesting Structures

Harvesting of local rainwater and recycling it for life saving irrigation is an old concept in India. Farmers have evolved different water harvesting system depending on rainfall and soil type. In medium to high rainfall areas, earthen embankments/farm ponds/Johad/stone checks are indigenously constructed by the farmers. These structures are also constructed for livestock, drinking and for recharging of the wells.

Dry Stone Masonry Pond/LCWHS

Dry stone masonry pond/LCWHS of 1m, 2m and 2.5 m height are constructed by the farmers to harvest rainwater. In this type of pond, the upstream and downstream walls are made up of dry stone and murrum/soil is filled in between them at an appropriate interval to provide stability to the structure. The design dimensions of upstream wall and downstream walls are taken the same. Dry stone masonry ponds should not be constructed for the height more than 2.5 m, over it the structure may overturn due to water pressure.

Cement Masonry Pond/LCWHS

Cement masonry pond/LCWHS upto height of 3 m is indigenously constructed by the farmers. In this type of pond, the upstream wall is constructed with cement masonry, whereas downstream wall is constructed with dry stone masonry. In between two walls murrum/soil is filled at an appropriate interval to provide stability to the structure. The design dimensions of upstream and downstream walls are taken according to the height of the structure but in any case, the bottom width should not be more than 1.5 m. The cost of different water harvesting structures, upto height of 2.5 m is given in the following Table:

Cost Comparison of Different Water Harvesting Structure

Height (m)	Cost per running meter of crest length (Rs) based on 1998 cost			
	Earthen Embankment (Johad)	Dry Stone Masonry Pond	Cement Masonry Pond	Anicut
1.0	142	416	793	3400
2.0	469	1030	1920	6200
2.5	745	1420	2700	8400

Sub-surface Harvesting

Sub-surface barriers are used to retain seasonal sub-surface flows and facilitate the abstraction of water through lined, shallow wells. The objective is to place an impermeable barrier–either of clay or masonry across the river bed from the surface down to an impermeable layer below. Another type, the raised-sand dam, can be constructed with masonry. This cam actually raises the level of the river bed, catching sand from the upstream and, in the process creating the groundwater reservoir. The dam is raised in successive stages, each stage of 50 cm, to achieve this effect. The clay dam is best built where a layer of clay underlies the sandy river bed. A masonry dam can be built where a bedrock layer occurs between 2 and 4 metres beneath the surface of the river bed. The raised-sand dam can be built at a point where the bedrock is exposed, or very near to the river surface, and raised in 50 cm increments after each rainy season.

Conclusion

India is facing a serious problem of water scarcity because of over exploitation of the existing limited water resources. Harnessing of rain by harvesting it suitably and increasing the recharging possibilities of groundwater could only be the befitting answer in negotiating with the recurrant droughts. There is an urgent need, therefore, to revive the traditional and the existing water harvesting system in various ecological regions of the country. Water harvesting must be made a national movement by creating it mandatory in rural as well as in urban areas and must be taken on community basis.

References

1. Anonymous, 1998. Hand Book on Rainwater Harvesting. Rajiv Gandhi National Drinking Water Mission, Ministry of Rural Areas & Employment, Govt. of India.
2. Sinha, I.N., 1997. Water conservation and management coalescence in drought mitigation. Water and Energy Environment, Vol. 54 (1): pp. 21-27.
3. Singh, P.K., 2000. Watershed Management-Design and Practices. E-media publication, Udaipur (India).

❑❑❑

Chapter 8

Rainwater Management for Enhancing Crop Productivity through Dovetailing Activities and Participatory Mode

D.H. Ranade

Operational Research Project for Dryland Agriculture, RVSKVV Campus, College of Agriculture, Indore - 452001 (M.P.)

Introduction

Besides many mandates, the Operational Research Project for the Dryland Agriculture, College of Agriculture, RVSKVV Campus, Indore has been actively working in different watersheds of Malwa region to achieve a firsthand working experience in the development of area on watershed basis to serve as a model for extension agencies and to provide consultancy services to the extension agencies for transfer of dryland technologies. It was observed through the project that in Malwa region, limiting moisture in the soil has been the major cause of concern like most of the rainfed areas due to uneven rainfall distribution, prolonged dry spells and vagaries of monsoon for growing satisfactory Kharif crops in general (Agrawal and Narain, 1997, Ariyabandu, 2001 and Rana and Gupta, 2010). Therefore, it is desirable to make the provision for sufficient irrigation water for this purpose either as surface and sub-surface water sources. Thus, dovetailing of various schemes and participation by the farmer through land, labour or cost can easily help to develop a water storage body of sufficient size which not only retained huge amount of runoff but also recharged the nearby open wells and tubewells. With this objective in view, the present study was made to explore such site, develop it so that both the purpose of surface storage and ground water recharge can be obtained with an aim to provide sufficient water to the farmers so that they continue to grow satisfactory crops even despite aberrant monsoon conditions by involving Operational Research Project for the Dryland Agriculture, College of Agriculture, RVSKVV Campus, Indore, and Department of Agriculture and farmers welfare, Madhya Pradesh.

Materials and Methods

In the year 2010, as per the requirement, under Operational Research Project for the Dryland Agriculture (ORP), a village Gaddukhedi (22°48′N and 75°50′E, altitude 540 m) was finalized by considering various aspects and in consultation with Administrative and Agriculture officials of Dewas district of Madhya Pradesh. After making several visits, a suitable ORP site is selected. Before formulating the technical programme, deciding research strategies, conservation approaches and technologies, it is desirable to collect basic information on various aspects.. The idea was to dovetail the activities in the village Gaddukhedi atleast for some development work. Based on the information gathered, a suitable site was selected and converted into a huge water storage body through the active participation by Operational Research Project for the Dryland Agriculture, Department of Farmers Welfare and Agricultural Development and the farmers of the area.

Results and discussion

Outcome of Participatory Rural Appraisal (PRA)

Through active participation of all the stakeholder, research and developmental agencies in the PRA and through secondary data, it was observed that the Vertisols of the village are low in Nitrogen content with medium P and high K level. The soil is very shallow near the ridge line and in the middle reaches; soils are medium to deep in nature. Further, it has been classified as rainfed agro ecosystem and oil seed based production system. The average rainfall is 1067 mm with average 45 rainy days. Thus, it offers potential (apart from the above identified need) for rainwater harvesting in the district. The runoff potential in this village is very high due to their low infiltration rate. The safe disposal/storage of this runoff would be beneficial. The storage of the runoff water would help the farmer to utilize it during the moisture stress period. Since the soil profile is underlain by basaltic murrum, even the stored water would no longer be retained for longer time. But its percolation may enhance or recharge the underground tubewell or open well available in the vicinity (Auja, 2007 and Naik, 2009). Even in village Gaddukhedi, due to poor/non-availability of groundwater, lot of area could not be brought under cultivation during rabi season many a times in this village. Since runoff and associated soil losses thereby reduced fertility, non-availability of surface as well as sub-surface water, limiting moisture conditions has been the major problems of the region, it was thought that attempts should be made to water availability in the farmers fields so that besides kharif crops i.e. mainly Soyabean, Maize, other rabi crops like Chickpea and Wheat are grown without any fail atleast in few area.

Water recharging cum harvesting tank

A suitable site in the field was selected considering all the hydrological aspects in April 2010 during the PRA exercise. In this portion, technical guidance was provided to line department personnel and an excavated tank was developed using watershed funds available with line department and finally a suitable cemented structure in the form of outlet was constructed totally under the guidance of ORP team in June 2010. This tank of 1500 cu.m. capacity was allowed to fill with the runoff water in the following monsoon season and as expected most of the water percolated and very limited water was available for the irrigation purpose as not much back water can be stored on the upward sloping upstream site.

Further, it was then decided to make use available of the cemented outlet waste weir during March 2011. For this purpose, the existing nala portion other than the tank area towards upstream was excavated using the back hoe loader machines so that the back water retained in the nala and may aid to nearby open well. Participatory mode was adopted for this purpose and farmers was asked to bear the cost of excavation by providing his tractor and trolleys so that the excavated soil is spread in the undulating fields of the farmer's fields and the basaltic material is used for the construction of field bunds. Also the excavated basaltic material was used at the field bunds for providing waste weirs in the bund portion so that the silt free water from the cultivated fields enters into the nala. This brought a sizable portion of the farmer's fields into levelled condition and also provided a safe disposal to runoff water.

However, in a watershed area, it is not possible to assess the direct or indirect advantage of this water harvesting cum recharge tank on many fields and farmers away from the tank area. Therefore, it was decided that the data are generated from a farmer who is directly drawing water from this tank area and from his open well located very nearby hardly 3 m and connected through tank with a pipe for recharging purpose. Thus, this study was carried out 2010 onward with an objective to generate data on water availability and to observe its effect on increase in Rabi cropped area.

From ORP side, the advantage of this outlet was explored and through this sunken pond, the nearby open well gets recharged automatically in the monsoon season of 2011-12. This helped the farmers to bring an appreciable amount of cultivated area under additional irrigation. Similarly, in the year 2012-13, the advantage of the increased storage capacity was very much visible, as the open well located in the adjoining field remained fully recharged and allowed farmer to draw more water than other wells located in the same area but away from the influential zone of the water harvesting tank. Similarly, during 2013-14, the heavy runoff due to more rainfall resulted in the tank remained full to its capcity almost throughout the monsoon period. The adjoining wells and tubewells got recharged and provided extra water available for irrigation. The farmer utilized the stored surface water by using floating submersible pump during the month of October and November to provide pre-sowing irrigation to adjoining fields and brought the fields under cultivation of gram and wheat. It is to emphasis that the site was such selected that despite erratic rainfall pattern, the tank area and extended portion remained full throughout the rainy seasons of 2010, 2011, 2012 and 2013 and provided surface irrigation water for pre-sowing irrigation in 10 ha area.

Conclusion

Thus, it can be concluded that dovetailing of various schemes and participation by the farmer through land, labour or cost easily helped to develop a water storage body of sufficient size which not only retained huge amount of runoff but also recharged the nearby open well and tubewell. Further, this has reduced the amount of runoff which was otherwise leaving the watershed area without aiding to surface and sub–surface storage. The increased water availability also brought new area under high remunerative rabi crops and increased their productivity on sustainable basis.

References

1. AISLUS (All India Soil and Land Use Survey) (1990). Watershed Atlas of India. Ministry of Agriculture, Government of India, New Delhi.

2. Ali, Shakir, Sethy, B.K., Singh, R. K, Parandiyal, A.K., and Kumar, A. (2017). Quantification of hydrologic response of staggered contour trenching for horti-pastoral landuse system in small ravine watersheds: a paired watershed approach. Land Degradation and Development, 28(4): 1237-1252. DOI: 10.1002/ldr.2517.

3. FAO (Food and Agriculture Organization) (1993). Land and Water Bulletin No. 1 Proc. Informal Workshop on Land and Water Integration and River Basin Mgmt., FAO, Rome.

❑❑❑

Chapter 9

Groundwater Recharge - A Bone for Scarce Zone

Dr. M.S. Pendke

Agricultural Engineer, AICRP for Dryland, VNMKV, Parbhani, India

Introduction

Water is a prime natural resource. Assured crop production in rainfed area can only be achieved, if supplemental irrigation can be provided. Groundwater is clearly the preferred source of irrigation for farmers. The Groundwater levels in the region had reached alarming levels early this year itself. Hence, attention is needed to enhance the groundwater potential through artificial recharging techniques. Recharging of borewells using designed filtration system was studied since 2014 and 2015 on 51 locations in assured rainfall zone of Marathwada region. India is the world's largest groundwater user. Over three, fourths of food grains production coming from the irrigated lands is contributed by lands irrigated by groundwater. The depletion of groundwater resource is a matter of great concern for human society. Large-scale pumping out of groundwater and negligible recharging has created 'water havoc' in these borewell-fed areas. The Groundwater levels in the region had reached alarming levels early this year itself.

Climate of the region

The average annual rainfall of the region is 889 mm and falls under assured rainfall zone. The rainfall is uneven and varies from year to year. The experiment was conducted at the experimental field of Vasantrao Naik Marathwada Krishi Vidyapeeth, Parbhani. Geographically Parbhani is located in between $19^{0}16'$N latitude and 76^{0} $47'$E longitude and at 409 m above mean sea level.

Open well recharge technique : Experimental setup

The filter unit was designed and constructed near the open well at demonstration field. The filter unit consists of three blocks. One is called primary filter with dimension as 0.6mx0.6m containing layers of stones, sand and metals. The primary filter collects major sediments from runoff water. Primary filter is joined by 4" dia. pipe to silt trapping unit (energy dissipation unit) which have rectangular notch to pass water to main filtration unit. The dimensions of second block are 1.1 m x 1m x 1.5m and third block act as a main filtration tank having dimensions as 2m x 2m x 2m with three layers of different filter material with. The layers are, such as 30 cm sand, 30 cm metal and 30 cm stones which act as filter material. The main filter unit is joined to open well by the 4" dia. PVC pipe. The filter unit is constructed with brick walls and cement concrete.

The suspended load material flows with surface runoff that flows through trench and collected to the silt trapping unit with 6" diameter pipe. To determine the filtration efficiency of filter system, the water samples at inlet were collected at various time interval. These samples were allowed to settle for 24 hours and then placed in hot air oven for 24 hours at 105^{0}C. After 24 hours the weight of dry silt is weighted for determination of silt content at inlet of filter. Similarly, the water sample at outlet i.e. in the well were also collected and followed the same procedure for analysis of silt content at outlet water. The difference.

Filtration efficiency

The filtration efficiency was found out by using formula suggested by Gideon, (1982).

$$\text{F.E.} = 100 \times 1 - \left(\frac{So}{Si} \right)$$

Where,

F.E. = Filtration efficiency (%)

So = Component concentration at the filter outlet (mg/lit)

Si = Component concentration at the filter inlet (mg/lit)

Specific gravity of soil solids

As the specific gravity of soil is the basic property and primary requisite for determination of velocities of silt contained water, it has been determined by standard pycnometer test.

Table 1: Specific gravity computation

	Observations
$M_{1(g)}$	380g
$M_{2(g)}$	636g
$M_{3(g)}$	1388g
$M_{4(g)}$	1228g

Specific gravity calculated as follows :

M_d=256g

Mass of water in (3) =752g

Mass of water in (4) =908g

Then,

$$G = \frac{256}{256-(1388-1228)} = 2.66$$

The specific gravity was determined and found to be 2.66 for collected soil samples.

Verification of Design criteria

Velocity of suspended soil solids

The soil samples collected from field through which runoff is flowing were analyzed for determination of particle size using particle size distribution curve. Accordingly, the particle size is found to be in the tune of 0.01 to 0.1mm.

The velocity at which muddy water flowing through the filtration process was computed by using stokes law,

$$\left(V = \frac{1}{18} D^2 (G-1) \gamma w / \eta \right)$$

Considering, determined specific gravity and standard values of viscosity and specific weight,

G=2.66

η=0.01 poise=10^{-6} kN-s/m^2

$$V = \frac{D2}{1.835} X(2.66-1)/10^6 \times 10^{-6}$$

V = 1.66 D^2 / 1.835

V = 0.9046 D^2

1. For coarsest particle D=0.1mm

   Vmax= 0.9046x$(0.1)^2$

   Vmax=0.9046X10^{-2} m/s

2. For finest particle D=0.01mm

   Vmin= 0.9046x$(0.01)^2$

   Vmin=0.9046x10^{-4} m/s

   V max and V min are 9.046x 10^{-3}m/s and 9. 046x10^{-5}m/s

$$Vavg. = \frac{(904.6x10^{-5} + 9.04x10^{-5})}{2} = 456.82x10^{-5} m/s.$$

The minimum and maximum values of velocity were determined as 9.046 x 10-3 m/s and 9. 046x10-5m/s with an average velocity value 456.82 x 10-5 m/s.

Standardization of filter material criteria

Alternatively the time required to flow water from inlet to outlet was noted which is 263 seconds and accordingly considering the average velocity value.

The head through which water passes was determined as:

h=v/t

h=456.82x10^{-5}/263

h=1.2 m=120 cm.

The head comes out to be 120 cm. Based on these calculations, the depth of filtration material was worked out and determined as 90cm keeping 30cm as a head of standing water which is essentially required for easy flow of water through filtration unit.

Table 2 : Specification of Filter material

Sr. No.	Filter material	Size	Thickness of layer
1.	Sand	0.4"	30 cm
2.	Metal	1.5"	30 cm
3.	Stones	3 to 4"	30 cm

Filtration efficiency of primary filter

The filtration efficiency of primary filter is found to be in the range of 64 to 70 per cent proves that it is capable of arresting the major silt component of the runoff water.

Table 3: Filtration efficiency of primary filter

Sample No.	Inlet silt composition	Outlet silt composition	Efficiency (%)
1	2.6	0.9	65.40
2	2.5	0.75	70.00
3	2.9	0.95	67.20
4	2.8	1.0	64.30
5	3.0	1.1	63.30
	2.76	**0.94**	**67.00**

Number of water samples were collected and analyzed to determine filtration efficiency of primary filter. Accordingly, the filtration efficiency of primary filter is found to be in the range of 64 to 70 per cent.

Based on the results filtration efficiency of primary filter, it is proved that primary filter is necessary to arrest the major silt component of the runoff water before it entered to the main filter unit. Primary filter helps to increase the life of main filter unit.

Filtration efficiency of main filter

The filtration efficiency of main filtration unit was determined in last three years based on the sample data collected from time-to-time interval. The year-wise data is presented in Table 4.

Table 4 : Filtration efficiency (2011-13)

Test Sample No.	Inlet silt composition (g)			Outlet silt composition (g)			Filtration Efficiency(%)		
	2011	**2012**	**2013**	**2011**	**2012**	**2013**	**2011**	**2012**	**2013**
1	2.8	2.9	2.5	0.2	0.2	0.13	92.85	93.10	94.80
2	3.2	3.1	2.9	0.2	0.2	0.2	93.92	93.54	93.10
3	1.5	3.4	3.3	0.1	0.12	0.3	93.3	96.47	90.90
4	3.5	3.3	1.5	0.3	0.15	0.1	91.42	95.45	93.33
5	2.5	2.2	2.8	0.15	0.15	0.2	94.10	93.18	92.85
6	1.0	2.3	1.6	0.12	0.12	0.14	90.10	94.78	91.25
					Average		**92.58**	**94.61**	**92.70**

The filtration efficiency is found to be in the range 92 to 95 per cent. The efficiency of main filtration unit is more than 90 per cent indicating that the performance of filter material with respect to its depth and quality.

Effect of well recharging on groundwater level fluctuation

The ground water levels (MSL) in well were recorded since 2011 to 2013 and the data is presented in Table 5. The water table starts rising since June after onset of monsoon. The increase in water table was observed upto the month of November and later a gradual reduction in water table was observed. The total runoff water from 1.8 ha field was diverted to well through well recharging model.

Table 5 : water level in well (M.S.L.)

Month	2011	2012	2013	Increase in water level in 2013 compared to 2011
January	404.8	405.0	405.6	0.8
February	404.6	404.5	405.2	0.6
March	404.6	404.1	404.8	0.2
April	404.0	403.9	404.4	0.4
May	403.9	403.3	404.2	0.3
June	404.0	404.6	407.4	3.4
July	405.0	405.5	407.7	2.7
August	405.2	406.1	407.7	2.5
September	406.7	406.2	407.7	1.0
October	406.3	406.2	407.6	1.3
November	406.8	406.9	407.7	0.9
December	406.7	406.5	407.5	0.8

During the year 2013, due to early onset of monsoon, the water table in well starts rising since May. Due to sufficient rainfall in the month of June, sudden rise in water table was observed. A well distributed rainfall was observed in monsoon season of 2013, a continuous higher water table was observed and upto November 2013, no reduction in water level was observed.

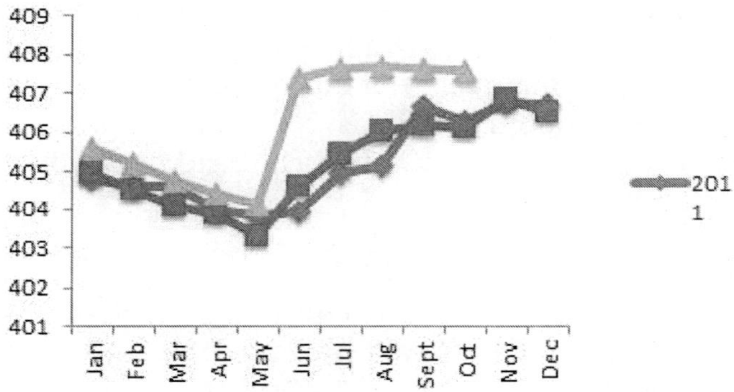

Fig. 2: Water Table (mm) as influenced by well recharging

Comparison of water levels of 2011 to 2013 indicated that there is an increase in water level in the tune of 0.3 m to 3.4 m. Thus, artificial well recharging on large scale can result in increasing the ground water potential so that supplemental irrigation to kharif and rabi rainfed crops can be provided.

Borewell Recharge Technique : Design Considerations

Using particle size distribution curve, the particle size is found to be in the tune of 0.01 to 0.1mm. The velocity at which muddy water flowing through the filtration process was computed by using stokes law,

$$V = \frac{1}{18} D^2 (G-1) \gamma w / \eta$$

Considering determined specific gravity and standard values of viscosity and specific weight.

G=2.66, η=0.01 poise=10^{-6} kN-s/m^2

$$V = \frac{D2}{1.835} X(2.66-1)/10^6 \times 10^{-6}$$

$V = 1.66\ D^2 / 1.835\ V = 0.9046\ D^2$

1. For coarsest particle D=0.1mm, $V_{max} = 0.9046x(0.1)^2$
   $V_{max}=0.9046X10^{-2}$ m/s
2. For finest particle D=0.01mm, $V_{min} = 0.9046x(0.01)^2$
   $V_{min} = 0.9046x10^{-4}$ m/s

The minimum and maximum values of velocity were determine as 8.228×10^{-3} m/s and $8.228x10^{-5}$m/s with an average velocity value 456.82×10^{-5} m/s. Alternatively the time required to flow water from inlet to outlet was noted which is 263 seconds and accordingly considering the average velocity value. The head through which water passes was determined as,

$h=v \times t$, $h=456.82 \times 10^{-5} \times 395$ h=1.8 m = 180 cm.

The head comes out to be 180 cm. Based on these calculations, the depth of filtration material was worked out as 150 cm with 30 cm as a head of standing water which is essentially required for easy flow of water through filtration unit.

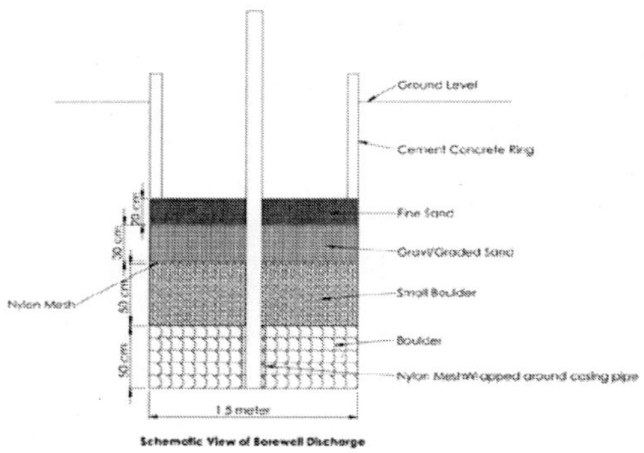

Fig. 3: Schematic view of Borewell recharge technology

Working principle of Borewell recharging system

Runoff water allows entering in primary filter unit, wherein the major sediments were arrested and water flows to the secondary filter unit. Secondary filter unit consist of excavation of soil around the bore well casing pipe by 2.5 m depth and 1.5 dia. From the bottom, up to 50 cm height, small holes are made with pointer at a spacing of 5 cm and this casing pipe is wrapped with nylon mesh. Then, the pit is filled with 4 layers of big stone, metal, gravel sand and fine sand one above each. On the top, the unit is covered with cement ring for not allowing the sediment from the flowing water.

Determination of Aquifer Parameters by Pumping Test

Papadopulos and Cooper (1967) curve-matching technique was adopted for determining aquifer properties. A long duration pumping test was conducted on representative borewell. The well characteristics like specific yield and transmissivity were determined as 0.0134 and 572.37 m^2/day respectively.

The filtration efficiency of the developed filtration unit of bore-well recharge model was found to be in the range of 93.22 to 96.47 percent with an average of 94.62 per cent.

Effect of well recharging on groundwater level fluctuation

The ground water levels in different wells were recorded before and after monsoon season of 2015 and the data is presented in Table 6.

Table 6 : Water levels and groundwater recharge in Borewells

Well number	Water level from GL in May 2015(Pre-monsoon)	Water level from GL in Nov. 2015 (Post Monsoon)	Water level fluctuation, m	Ground water recharge, cm	Ground water recharge %
	TREATED BOREWELLS				
1	31.45	23.20	8.25	11.055	27.16
2	17.27	10.05	7.22	09.675	23.77
3	18.55	12.23	6.32	08.468	20.80
4	32.69	23.85	8.84	11.845	29.10
5	33.12	24.51	8.61	11.537	28.34
6	14.34	8.33	6.01	8.053	19.78
7	21.72	15.09	6.63	8.884	21.82
8	15.85	9.72	6.13	8.214	20.18
9	19.20	12.35	6.85	9.179	22.55
10	72.25	66.23	6.02	8.066	19.81
11	40.60	34.70	5.90	7.906	**22.26**
12	26.85	19.60	7.25	9.715	**23.86**
				Average	**23.28**

UNTREATED BORE WELLS					
13	74.50	71.18	3.32	4.44	**05.94**
14	29.50	26.61	2.89	3.87	**05.18**
				Average	**05.56**

Data revealed that the water table starts rising since September. The increase in water table was observed upto the month of November and later on gradual reduction in water table was observed.The specific yield value was used for estimation of ground water recharge from the borewell based on water table fluctuations in the borewells. It was observed that in recharged borewells, the water table fluctuation was in the range of 5.90 to 8.84 m. It was also revealed that the groundwater recharge in treated borewell was in the tune of 19.78 % to 29.10 % with an average of 23.28 % as against average recharge of 5.56 % in untreated borewells. The water levels in recharged and un-recharged borewells were recorded in May 2015 and May 2016.

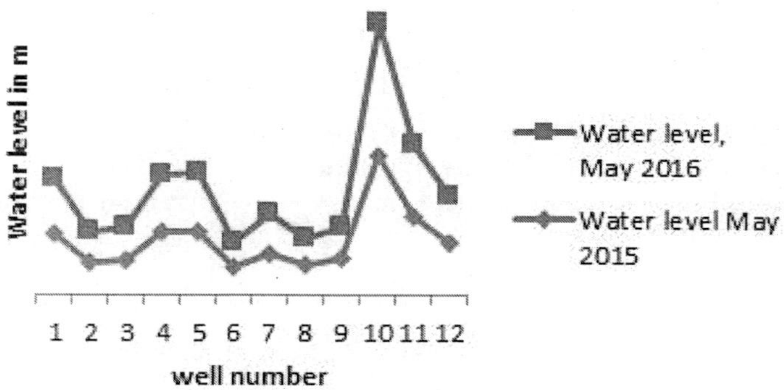

Fig..4: Comparison of water levels in recharged borewells in May 2015 and May 2016

It is found that the water level is increased in the tune of 1.87 to 3.30 m with an average of 2.37 m in the zone. However, in un-recharged borewells, the water levels was decreased by 0.62 m in 2016 as compared to 2015

Conclusions

1. The filtration efficiency of the designed filter was found to be 93 to 96 %.

2. The water level fluctuation in pre-monsoon and post-monsoon season in recharged bore are found to be in the range of 5.90 to 8.84 m.
3. The groundwater recharge in treated borewell was found to be 23.28 per cent of annual rainfall as against 5.56 per cent in untreated borewells.
4. The rise in water level in recharged borewell is found to be 2.37 m as compared to fall of 0.62 m water level in un-recharged borewell during May 2015 to May 2016.

References

1. Aggrawal and Soni (2005), Study of various types of filter in roof-tops rainwater harvesting, proceeding 38th annual convention of Indian Society of Agricultural Engineers, pp. 75-82.
2. Bhalerao S.A. and T.S.Kelkar(2013) Artificial recharge of groundwater : A novel technique for replenishment of an aquifer with water from the land surface, International journal of Geology, earth and environmental science, 3(1): 165-183.
3. Mehta, M.S., Murwaha, D.C., Sharma and Dalal Singh (2002). Artificial recharge to groundwater through roof top rain water harvesting. A case study of union territory of Chandigarh. All India Seminar on Artificial recharge of groundwater, pp. V45-V49.
4. Pratima Patel, Mahesh Desai and Jatin Desai (2011) Geotechnical parameters impact on artificial groundwater recharging techniques for urban centres; Journal of water resources and protection vol13, pp. 275-282.
5. Punmia, B.C. (1973). A textbook of soil mechanics, Laxmi Publication Pvt. Ltd., New Delhi pp. 179 and pp. 41-42.
6. Ravichandran S.S. Sathish Kumar and Leena Singh (2011) Selective techniques in artificial groundwater recharge through dugwell and injection well methods; International journal of chemtech research, 3(3): 1050-1053.

❑❑❑

Chapter 10

Artificial Groundwater Recharging : A Bone for Scarce Zone

Shakir Ali

ICAR-Indian Institute of Soil and Water Conservation (IISWC), Research Centre, Kota-324002 (Rajasthan)

Introduction

Water is lifeline of all plants and animals on the earth. Water is needed to ensure food security, feed livestock, maintain organic life, take up industrial production and to conserve the biodiversity and environment. Hence, there is no life without water. The Earth is the only plant, so far known to have water and this makes it fit for human living. Nearly 70% of the earth surface is covered with water, which amounts to 1400 million cubic kilometres (m km3). However, 97% (1320 m km3) of this water being sea water, it is salty. Fresh water availability is only 3% (37.5 m km3) and only 40% of this can be used by human beings. Out of the total fresh water, 77% is frozen in ice glaciers/caps, 22% is stored underground and only 1% water is available on the surface of the earth. Out of the surface water, 87% is stored in lakes, 11% in swamp and 2% in rivers (Anon. 2006). In India, the per capita water availability in 1951 and 1955 was respectively 5177 and 4732 m3 per year, when the total population was only 361 and 395 million, respectively. In 1991 and 2001, as the population were increased to 846 and 1027 million, the per capita water availability reduced drastically to 2209 and 1820 m3 per year, respectively.

Out of the total cultivable area of 182 m ha, only 140 m ha (77%) are under net cultivation and of this, 62 m ha (44%) are under irrigation. There is further potential to increase the area under irrigation to 140 m ha, 76 m ha (54%) through surface water and 64 m ha(46%) by using groundwater. So far, the irrigation potentials have already been created to cover 107 m ha, although they are not utilized effectively. It is estimated that effective area under irrigation by 2025 will be 76 million ha, although the Government of India is estimating to cover 104 million ha. Groundwater is the major source of irrigation and this trend will continue. By 2025, 60 million ha will be irrigated by using groundwater and by 2050, the area underground water will increase to 70 million ha. In 2000, the area under canal irrigation was 17 million ha, which will increase to 27 million ha by 2050. There is further scope to increase the potential by 35 million ha, by inter-linking the rivers and harnessing 36 billion m3 through artificial recharging of groundwater (Government of India, 2009). Indian's groundwater system is the basic resource that fulfills 70% irrigation water requirements including the fruit crop production (MOA, 2013; CGWB, 2014). Due to increasing demands for expansion of agriculture and growing industrialization, over exploitation, continuous failure of monsoon and non-adoption of manage aquifer recharge (MAR) facilities, there are decline of water table and deterioration of water quality of the groundwater in many states of India. The water table decline has been resulted in drying up shallow wells, low yields and higher pumping costs. As per the assessment made by Central Groundwater Board Ministry of Water Resources, River Development & Ganga Rejuvenation in 2014, out of 6607 of assessed administrative units, 1071 units are over-exploited, 217 units are critical, 697 units are semi-critical, and 4530 units are safe. Apart from these, there are 92 assessment units which are completely saline. Over-exploited zone denotes that the stage of groundwater development (SGWD) is more than 100% of net annual groundwater availability with significant decline in water table. The SGWD between 90 and 100%, and between 70 and 90% of net annual groundwater availability with significant decline water table are called critical and semi-critical zone. In case of safe zone, SGWD is less than 70% of net annual groundwater availability and there is no decline in long-term water level. The over-exploited states are in order of Punjab (172%) > Rajasthan (137%) >Delhi(137%) >Haryana(133%). The critical stages are UTs of Daman & Diu(97%), and Puducherry (90%). In the states of Tamil Nadu(77%), Uttar Pradesh(74%), and Himachal Pradesh (71%) are under semi-critical stage. The rest of the states/UTs are under safe zone.

Groundwater recharging facilities

Groundwater recharge is considered to occur via different pathways in arid and semi-arid regions. First natural recharge, in which, rainfall infiltrates vertically into soil and hits on water table to aquifer, called natural/diffused recharge. Second enhance recharge, which distinguished from natural recharge, which is defined as the natural recharge by man-made changes/modification in land surface, landuse system and vegetation to increase natural recharge is called enhance recharge. Natural recharge is expected to be much slower than artificial recharge due to low pressure head on surface, however, it is significant component of watershed's water balance, if rate of recharge is non-zero over extensive area. Third incidental recharge, which is defined as recharge from human made activity or natural not designed specifically for recharge. It includes recharge from irrigation field, septic tank, leach field, storm water infiltration structures waste water facilities, canal, stream and rivers. Fourth induced recharge, in which infiltration wells/galleries are placed near the river bank or water body. When these wells pump groundwater, it induced flow out from river/body into infiltration wells. Fifth artificial recharge, it is a type of controlled recharge, where surface water is put on or in the ground for infiltration and subsequent movement to the aquifer to augment the groundwater resources. It can be defined as the methods of increasing the groundwater of the aquifer by artificial means as engineered structures. The AGR occurs via infiltration or injection.

Recharge Filter for artificial recharging of dug/tube well

India has 19.86 million wells which includes dugwells (0.92), and shallow (9.12) and deep (1.54 million) tubewell and probably rank first in the World (Shah et al., 2003; MWRRDGR, 2014). Out of which, 11.72 wells are in use and rest are in continuous uses (Mohan, 2016). There is declined 4.32 % in dugwell, and increases in shallow and deep tube by 9.22 and 171.70% in 2004 over the 2001 estimates. Total irrigation potential created by these wells is 7.78 m ha. In states of Rajasthan, have the highest number of wells followed by Tamil Nadu, Haryana, Maharashtra, MP and undivided AP Undivided Andhra Pradesh. These six states together account for a major share of the country's wells as 96.9% of total wells and 96.57% of wells that are in-use. In case of dugwells, Maharashtra is the leading state, whereas Uttar Pradesh and Punjab are leading in tubewells (shallow and deep respectively). Recharge filter is one of an efficient and cost effective AGR method to increase groundwater storage capacity functional and/or abandoned dug and tubewells.

A. Dugwell Recharging

Recharge filter for the dugwell is composed of a filtration unit near dugwell filled up with filter materials in layers. The filled material are usually in three layers with the pebble, P (20-30mm); gravel, G (5-20mm); and course sand, CS (0.2-2mm) at the bottom, middle and top of the recharge filter, respectively which acts as fitter. The ratio of the thickness of filtration layers is usually kept 1.5:1:3(CS: G: P) to facilitate higher recharge and good water quality (Kambale et al., 2009; Kumar et al., 2012). The nylon net is spread over the gravel layer to separate it from sand layer and to prevent silt particles filling the void spaces in lower layers. The nylon net helps to allow only clean water to pass through it. A PVC pipe of 100-150mm diameter with a wire mesh filter is fixed at the bottom of the filtration unit which opens into the dugwell. The width of recharge filter with rectangular cross section were 1.25, 1.50, 1.75, and 2.0 length-width ratios are respectively, 1.14, 1.04, 0.96 and 0.90m, and corresponding length are 1.42, 1.56, 1.68 and 1.80m. There is a wide variation in optimum sizing of recharge filter for dugwell, which varied with each input parameters of the recharge filter.

B. Tubewell Recharging

Artificial groundwater Recharging of tubewell is very similar as dugwell. The designed size recharge pit is dug out around the tube well. When there is a chance of casing pipe slipping into the tubewell after the pit is dug, to avoid this danger, cement concrete should be laid around the casing pipe at the bottom of the pit and a clamped to fixe in the cement concrete to tightly hold the casing pipe from slipping into the tubewell. Using iron pipe as casing pipe causes problems like rusting in a few years, in such cases we should replace them with PVC pipe and make 300 to 400 holes of size 6mm with 3 inch distance each upto a length of 5 ft from the bottom of the pit. Slotted pipes available in the market can also use. These holes need to be covered with nylon mesh or aqua mesh and were tied tightly to the casing pipe to prevent the passage of silt and dust through the holes. Impact of artificial recharging of abandoned and/or functional tubewell through recharge filter in AP and Karnataka showed a multiple benefits. technology to an abandoned dugwell in the Dhoti watershed, Baran district, amount of harvested runoffs recharged to the defunct dugwell ranged between 1188 and 3499 m3/yr with a mean of 2343.5 m3/yr. The improved in the depth to water table ranged between 6 and 8m with a mean of 7m.

C. Recharge ponds/basin

Among the recharge facilities schemes, recharge pond/basin is the oldest and a much preferred and accepted technology in the India by the farming community due to its simplicity, low repair and maintenance cost and multiple services to local community. These ponds are commonly known as recharge or percolation pond or tank. In India, pond is known by different names in different States, viz. it is known by Johad, Nadi, Paal and Khadin in Rajasthan; by Tal and Bandhara in Maharastra; by Bundh in Madhya Pradesh and Uttar Pradesh; by Ahar and Pyne in Bihar; by Puzhas in Kerala, by Oorani in Andhra Pradesh and by Eris in Tamil Nadu (Ali et al., 2002). These ponds are small reservoir excavated out in a watershed or by small earthen dam constructed across the natural stream to catch and hold the surface runoff generated during the monsoon season. The stored runoff infiltrates naturally from the recharge pond to underneath aquifer. The size of these ponds varies between 0.2 and 1.0 ha, and depth ranges from 1.5 to 3m. A study in a BK watershed for an experimental recharge pond showed the potential recharge to the aquifer was in the range of 76.6 to 90.8% of the stored runoff. Nearly, 7.2 to 8.7% of water was lost through evaporation, 0 to 15.5% as outflow, and 0.6 to 1.2% remained as storage at the end of simulation periods.

D. Check dams

Masonry check dams (MCDs) usually constructed at the lowest reach of the main gully to retain silt free runoffs from the upper and middle reach of the watershed. The water retained in the check helps in facilitating live saving irrigation to the crops, artificial groundwater recharging and improving the groundwater level.

References

1. Ali, Shakir, Singh, R. K, and Sethy, B.K. (2015). Field application of a dynamic potential groundwater recharge simulation model for

small recharge ponds. Applied Engineering in Agriculture, ASAE, 31(2): 267-279. DOI 10 13031/aea.31.10516.

2. Anonymous, 2006. Water–A shared responsibility, United Nations World Water Development Report, 2.
3. CGWB (2010). Artificial Recharge to groundwater through dug wells (dugwell Recharge scheme). Central Ground Water Board (CGWB), GOI, New Delhi, India. Pp 29.
4. CGWB (2013). Master plan for artificial recharge to groundwater in India. Central Ground Water Board (CGWB), GOI, New Delhi, India. Pp 225.
5. DDWS (2009). Mid term appraisal of the 12th five year plan-rural water supply. Department of Drinking Water Supply (DDWS), Ministry of Water Resources, GoI, New Delhi.
6. Govt. of India, 2009. Background note for consultation meeting with Policy-makers on Review of National Water Policy. Ministry of Water Resources. 50 pp.
7. Kambale, J.B., Sarangi, A., Singh, D.K., and Singh A.K. (2009). Performance evaluation of filtration unit of groundwater recharge shaft: laboratory study. Curr. Sci., 96(4): 471-474.
8. Ministry of Agriculture, Government of India (2013). State of Indian Agriculture: 2012 13. Department of Agriculture and Co-operation, Ministry of Agriculture, Government of India, New Delhi. pp. 221.

□□□

Chapter 11

Success Stories towards Climate Resilient Agriculture through Participatory Adoption of Technologies and Mitigation Activities

A.K. Kothari, R.K. Sharma, S.K. Sharma, G.R. Chary, Ch Srinivasa Rao, J.K. Balyan and Devyani Dodia

All India Co-ordinated Reseach Project for Dryland Agriculture Dryland Farming Research Station, Bhilwara, (Maharana Pratap University of Agriculture and Technology, Udaipur)

Introduction

Climate change and variability are concerns of humankind. The recurrent droughts & floods, cold and heat waves threaten seriously the livelihood of billions of people who depend on land for most of their needs. Climate change has become an important area of concern for India to ensure food and nutritional security for growing population. The impacts of climate change are global, but countries like India are more vulnerable in view of the high population depending on agriculture. In India, significant negative impacts have been implied with medium-term (2010-2039) climate change, predicted to reduce yields by 4.5 to 9 per cent, depending on the magnitude and distribution of warming. Since agriculture makes up roughly 16 per cent of India's GDP, a 4.5 to 9 per cent negative impact on production implies a cost of climate change to be roughly upto 1.5 per cent of GDP per year. Therefore, to meet the challenges as posed by climate change on the agricultural system, the Government of India has accorded high priority on research and development to cope with climate change in agriculture sector and launched the project entitled "National Initiative on Climate Resilient Agriculture (NICRA) "in February, 2011, which is a network project of the Indian Council of Agricultural Research (ICAR). The project aims to enhance resilience of Indian agriculture to climate change and climate vulnerability through strategic research and technology demonstration. The research on adaptation and mitigation covers crops, livestock, fisheries and natural resource management. The project consists of four components viz. Strategic Research, Technology Demonstration, Capacity Building and Sponsored/Competitive Grants. The technology demonstration component of the NICRA project is covering 100 KVKs in 8 zones, ToT divisions of 7 core ICAR institutes and 23 centers of All India Co-ordinated Research Project on Dryland Agriculture network (AICRPDA). Under this component, an integrated package of proven technologies are being demonstrated in one village panchayat in each district for adaptation and mitigation of crop and livestock production systems to climate variability based on available technologies. The effect of climate change is much severs in resource poor rainfed area which occupies 68 per cent of country's cultivated area and supports 40 per cent of human and 60 per cent of livestock population.

The NICRA Project at Dryland Farming Research Station, Arjia, Bhilwara is one of the AICRPDA centers in India working for assessing the feasibility of technology adoption in dryland areas of south-east Rajasthan. The operational area of NICRA project of study is located in Bhilwara and Rajsamand district of Rajasthan (Table 1 & Fig. 1) are situated in zone IV-a of the Rajasthan and National Agro-Ecological Sub-Region IV. The zone is having predominant Maize based cropping system and in texture having shallow to medium depth. It receives an annual rainfall of 670 mm, out of which 90 per cent is received during June to September. The principal crops of the region are maize, groundnut, pulses & sorghum fodder during kharif and wheat, barley, mustard and chickpea during rabi. Two major locations were selected for the project activities. Out of this, the selected villages Kochariya and Mandpiya are situated in Bhilwara district ($24^0 20'$ N latitude and $74^0 20'$ E longitude) and located 20 km away from main centre and while the selected villages Lapsiya and Tara ka Hheda are situated in Rajsamand district (located between 24^0 46' to 26^0 01' N latitudes and 73^0 28' to 74^0 18' E longitudes) and 80 km away from Bhilwara i.e. Main Centre. Participatory Rural Appraisal, Focused Group Discussion and SWOT analysis were carried out to document the benchmark status of NICRA villages and to plan different activities for climate resilient agriculture. Demonstrations on rainfed technology were conducted under 4 sub-projects viz.; Real time contingency plan implemented on participatory mode, Rainwater harvesting (in-situ and ex-situ) and efficient use, Efficient energy use and management and Alternate landuse. The success of these activities has been compiled with an emphasis on participatory approach and lesson learned for further extension of the work.

Inadequate water resources make it imperative to evaluate the efficiency of water utilization to arrive at an efficient type of irrigation. "Water-use efficiency and productivity differ according to different systems of irrigation, crop mix and environment, and are comprised of different parameters: crop consumptive use (water requirement), an efficient crop mix (the maximum irrigable area for given water resources) and maximum output and value per unit of water." Various management practices and irrigation technologies are available to enhance efficiency of applied water in irrigated agriculture. Irrigation improvements often involve upgrades in physical application systems, with improved field application efficiencies and higher yield potential. Improved water management practices may also be required to achieve maximum potential of the physical system. In some cases, the effectiveness of improved irrigation practices may be enhanced, when implemented in combination with other farming practices, such as conservation tillage and nutrient management–good land husbandry is a pre-requisite for effective irrigated agriculture. Increasing irrigation efficiency would lead to 'saved water', which can either be used to augment crop yield (expand irrigated area), allow more water for downstream users or release water for other uses. In the light of climate change, improving irrigation efficiency is a plausible adaptive strategy. Therefore, an integrated and multidisciplinary research is required to develop operational technologies for mitigating climatic risks under inadequate water resource conditions.

SUCCESS STORY-1

1. Rainwater conservation through chiseling and peripheral bunding

In general, crop suffers due to poor germination and intermittent drought in the region. Therefore, in-situ rainwater conservation technology is the best bet to overcome the problem of germination due to limited moisture and to mitigate the intermittent drought in the region. It is a cost effective technique to improve rainwater use. Enhancement in-situ in rainwater conservation can be achieved by improving the opportunity time for infiltration through peripheral bunding and improving infiltration rate by breaking the plough pan. It is a preparatory technique for climate resilient agriculture.

Existing practice

Primary tillage is carried out by country plough/tractor drawn cultivator/MB plough/Disc plough for creating a rough soil surface but cultivating the land at a particular depth results in formation of plough pan which is the main reason for decrease in infiltration rate of rainwater in the soil profile. Further, farmers are not aware about improving the infiltration opportunity time for in-situ rainwater conservation.

Technological intervention

Chisel plough has been introduced to break the plough pan in the region and peripheral bunding/compartmental bunding has been introduced for improving the infiltration opportunity time for in-situ rainwater conservation.

Chiesel plough to break hard pan

Chiseling for in-situ rainwater management

Ridging after sowing (30DAS) in maize

Efficient tillage management for double cropping

Impact

On farm trials on in-situ moisture conservation in maize were conducted during 2011-12 to 2012-13 at farmer's field NICRA villages. The results revealed that in-situ moisture conservation in maize gave 18.53 per cent higher grain yield (2667 kg ha^{-1}) over the farmer practice (2250 kg ha^{-1}) with the highest net return (Rs 33750 ha^{-1}) and B:C ratio 3.75 over farmer practice (Rs 28133 ha^{-1}, B:C ratio 3.48).

Lesson learned

In-situ rainwater conservation can effectively be achieved by improving the availability of specific implement like chisel plough and peripheral/compartmental bunding.

SUCCESS STORY-2

2 Rainwater harvesting through farm Pond /Nadi

Rainwater harvesting is an utmost important technology for achieving climate resilience in dryland agriculture. The rainfall analysis in the region reveals that there is more than 75 per cent probability of dry spell during the terminal stage of major crops likes' maize and groundnut and on the other, there is surplus water available during the early and mid season of crop growth which can be harvested. Under such situation, Rainwater harvesting and recycling is only solution for improving the crop yields in the sustainable manner.

Existing practice

Maize and groundnut crop is predominantly grown in this region, which are usually suffer from moisture stress during maturity stage of crop growth and even farmers faces crop failure due to soil moisture stress and very rarely farmers able to sow rabi crops in the region, if sufficient late rain occurs.

Technological intervention

For rainwater harvesting, a basket of three technologies is available, which includes dugout farm pond with and without lining and Nadi-earthen embankment for surface storage (Nadi). These three options were given to farmers to select any of one technology and results are as follows:

Case A : Farm pond without lining

Smt. Kesar Devi w/o Sh. Badri Gurjar, Kochariya, P.S. Suwana, Distt. Bhilwara-Rajasthan has adopted Farm pond without lining (Kuccha) of size 1242 cum looking to her soil condition (Alkaline soil with murad at lower depths). The dimension of farm pond was 18 x 30 m at top, 12 x 24 m at bottom, 3 m deep having 1:1 side slope. The catchment area of the pond is 55 ha and it has a command area of 1.75 ha. After having a water source, she had change her cropping plan from maize and black gram during kharif and fellow during rabi to cotton, maize +black gram (2:2) intercropping and groundnut +sesame (62:2) intercropping during *Kharif* and mustard during rabi during the year 2012 and 2013.

Rainwater harvesting through Kuccha farm pond

Rainwater harvesting through Pucca farm pond

Case B: Farm pond with lining

Shri Kishan Lal S/o Shri Hema Jat, Bagatpura, panchayat-Lapsiya, P.S. Railmagra, District Rajsamand-Rajasthan had adopted a farm pond with cement concrete lining of size 240 CUM looking to his soil condition (loamy soil with fractured rock in depth) during the year 2012. The dimension of farm pond was 12 x 11.6 m at top, 4.5 x 4.2 m at bottom, 3 m deep having 1:1 side slope. Looking to soil condition it has been assumed that there will be a need of lining to control the seepage losses. Participatory discussion was made with the farmers and it was decided to go for cement concrete lining (1:4:8) of 5.0 cm thickness. The catchment area of the pond is 5 ha and it has a command area of 0.75 ha. He has also changed his cropping plan from sorghum and green gram during *kharif* and fellow during *rabi* to cotton, maize +black gram and groundnut +sesame during *Kharif*.

Rainwater harvesting through NADI

Pumping of harvested Rainwater

Case C: Nadi for surface storage of rainwater

Shri Shiv Singh S/o Sh. Ram Singh Shaktawat, village Mandpiya, Panchayat Kochariya, P.S. Suwana, Distt. Bhilwara-Rajasthan has adopted Nadi system modification for harvesting the rainwater of size 111375 cum looking to slope of the land and availability of present system. He had deepened the existing Nadi and modified the outlet for improving the recycling of harvested water in the year 2011-12. The length of the embankment was 55 metre with an average depth of 8.0 feet. After having such an assured water source, he had changed his cropping plan during rabi 2013 gram, was taken by the farmer.

Efficient utilization harvested Rainwater through sprinkler

Effect of harvested Rainwater application in Mustard

Effect of harvested Rainwater application in Gram

Impact

The optional technology of choosing a rainwater harvesting structure among Kuchha farm pond, Pucca farm pond having lining with polyfilm or cement concrete, modification of existing nadi, etc. as per site condition and farmer's preference.

Case A: Farm pond without lining

An on farm trial on rainwater harvesting and recycling was conducted at village Kochariya. The results revealed an increase of 132.53 and 169.09 per cent in net income (Rs. 35090 and Rs 49110) were recorded due to farm pond technology against a net traditional practice (Rs. 15090 and Rs. 18250/-) during the year 2012-13 and 2013-14. Therefore, it can be concluded that the cost of farm pond could be recovered within two years time.

Case B: Farm pond with lining

An on farm trial on rainwater harvesting and recycling was conducted at farmers field of village Bhagatpura. The results revealed an increase of 60.49 and 44.67 per cent in net income (Rs. 37010 and Rs. 44563) was recorded due to farm pond technology against a net traditional practice (Rs. 23060 and Rs. 18250/-) during the year 2012-13 and 2013-14. Therefore, it can be concluded that the cost of farm pond could be recovered within two years time.

Case C: Nadi for surface storage of rainwater

The results revealed a net income of Rs. 264500/- was recorded due to renovation of Nadi technology with a B:C ratio of 1.80. Therefore, it can be concluded that the cost of farm pond could be recovered within two years time.

Lesson learned

The farmer should be given a basket of technologies and as per his suitability according to local conditions for constructing a rainwater harvesting structure among different options like farm pond without lining, farm pond having lining with polyfilm or cement concrete or brick/stone slab lining, etc. and Nadi as per location characteristics and his preference. Inspite of this, there is a provision of subsidy for construction of dugout unlined farm pond of size 1200 cum as a blanket technology irrespective of land slope, soil condition, catchment area, farmer's preference, etc. by the Department of Agriculture, Government of Rajasthan, which may be modified in such a way that aforesaid optional technology can be promoted and facilitated to the farmers in the region inspite of a blanket technology.

SUCCESS STORY-3

3. Real time contingency planning through crops and its varieties

Successful crop production in dryland areas must make efficient use of precipitation that is often limited and erratic in spatial and temporal distribution. Soils of the area are light to medium textured with low available water holding capacity and fertility status. The major crops grown are maize, groundnut, blackgram, greengram, sesame and sorghum fodder during the *kharif*. The onset of monsoon is the only rainwater balance parameter known prior to major decision making for crop planning. Thus, the variability of onset of monsoon should be considered as key factor for dynamic selection of crops and its cultivar in different years to harness more profit per unit of rainfall under dryland conditions in semi-arid agro-eco region.

Existing practice

In general, farmers of the region have an affinity to the maize crop irrespective of time of onset of monsoon and they are pre-occupied as related to cropping plan of their farm. Besides this, farmers grow traditional varieties of mostly kharif crops. The process for change in crop planning due to change in climatic conditions is very slow till that they suffer a lot with respect to farm produce and economics.

Technological intervention

Trials and demonstrations have been conducted for performance evaluation of different varieties of different crops to suggest suitable crop and its cultivars as per real time contingency plan at farmer's field with local cultivar. Different improved varieties of maize viz., PM-3, PEHM-2, Aravali and Navjot; Sorghum varieties viz., CSV-15, CSV- 17 and Pratap 1430; blackgram varieties viz., T-9, TAU-1 and TAU-2 and groundnut varieties viz., TPG 41, TG 37A, TAG 24 and Pratap Mungfali 1 were tested in the region during 2011 to 2013.

Clusterbean varietal

Blackgram varietal

Sorghum varietal

Maize varietal

Taramira varietal

Impact

The village level climate facilitors become a key person in the village for disseminating the information right from movement of monsoon to district level forecast of climate scientific data. Because, farmers use this information for crop planning and manage the time of different agricultural operations according to district level climatic scientific data forecast. Three years study results revealed that maize variety PM-3 was found best suited in the region and gave higher yield 2547 kg ha^{-1} with higher net return (Rs.27192) and B: C ratio (3.15) as compared to traditional cultivar (1750 kg ha^{-1}, Rs.18432 and 2.56), respectively. Although, Maize variety PEHM-2 also found best suited for the region.

Three years study results revealed that blackgram variety T-9 was found best suited in the region and gave higher yield 463 kg ha^{-1} with higher net return (Rs.7850) and B: C ratio (1.66) as compared to traditional cultivar (279 kg ha^{-1}, Rs.-320 and 0.97), respectively. Three years study results revealed that groundnut variety TG 37 A was found best suited in the region and gave higher pod yield 1917 kg ha^{-1} with higher net return (Rs.41680) and B: C ratio (3.45) as compared to traditional cultivar (967 kg ha^{-1}, Rs.30147 and 2.75), respectively.

Lesson learned

Awareness among farmers about short duration improved cultivars of different cultivars and weather forecasting at village level well in advance will ensure to increase the productivity level and had great effect on economics.

SUCCESS STORY-4

4. Intercropping systems for drought proofing

Sowing of two or more crops in lines is termed as intercropping system. The crops are selected in such a manner that they significantly differ in their water requirement for insurance against aberrant to normal weather situation. The crops so selected should also be significantly differ in plant type specially plant height to efficiently utilize the sunlight by increasing the surface area. The intercropping system also increase the land equivalent ratio and existing practice.

Farmers used to grow two or more crops as an insurance against drought and to meet their food requirement. Mixed seeding of maize with blackgram is most common in this region. In groundnut, farmers broadcast sesame seed followed by planking. Whereas mixed seeding of sorghum with greengram is also most common practice in the sorghum growing NICRA villages.

Technological intervention

Intercropping of maize + blackgram (row ratio 2:2 at 30 cm spacing), intercropping of sorghum + greengram (row ratio 2:1 at 30 cm spacing), intercropping of groundnut + sesame (row ratio 6:2 at 30 cm spacing) and intercropping of blackgram + sesame (row ratio 2:2 at 30 cm spacing) are recommended against their respective mixed cropping. Further, there has been an acute need of mechanization due dwindling of bullock power resource for sowing. Therefore, suitable intercropping seed drills have been designed and introduced to improve the adoption in the NICRA village through custom hiring centre of improved farm implements.

Profitable inter-cropping system of maize + blackgram (2:2)

Profitable inter-cropping system of sorghum + greengram (2:1)

Most profitable inter-cropping system of groundnut+sesame (6:2)

Most enumerative inter-cropping system of maize+blackgram (2:2)

Intra-cropping of mustard across gram at 4m apart

Impact

Inter-cropping systems viz., maize + blackgram (2:2), groundnut + sesame (6:2), sorghum + greengram (2:1) and blackgram+ sesame (2:2) performed better at farmers field during drought. Results of three years on-farm assessment for suitability of maize + blackgram (2:2) inter-cropping system at farmers field revealed inter-cropping system of maize+ blackgram (2:2) gave 20.55 per cent higher maize equivalent yield over to mixed cropping maize (2365 kg ha^{-1}).

On-farm assessment for groundnut + sesame (6:2) inter-cropping system performed better at farmers field and gave 29.61 per cent higher groundnut pod equivalent yield over to mixed cropping groundnut and sesame (1155 kg ha^{-1}). Sorghum + greengram (2:1) inter-cropping system performed better at farmers field and gave 31.79 per cent higher groundnut pod equivalent yield over to mixed cropping groundnut and sesame (2139 kg ha^{-1}).

Whereas, results of three years on-farm assessment for suitability of blackgram+ sesame (2:2) inter-cropping system at farmers field revealed inter-cropping system of blackgram+ sesame (2:2) gave 41.77 per cent higher maize equivalent yield over to mixed cropping maize (328 kg ha^{-1}).

Lesson learned

Timely availability of inter-cropping seeed drill/drauft power and interest of farmers are the key factors for adoption of inter-cropping.

- Adoption of inter-cropping system can be improved by introduction of suitable intercropping seed drill.
- The use of intercropping seed drill should be promoted on community and sequence basis in order to judicious use of sowing time window.

SUCCESS STORY-5

5. Agro-advisories services for end users through ICTs

The Indian meteorological Department, Government of India is regularly releasing district level forecast of climatic parameters like rainfall, minimum and maximum temperature, minimum and maximum relative humidity, cloud cover (Octa) and wind speed and direction for consecutive five days to reduce the uncertainties for agricultural planning. The information so generated is useful to improve the productivity at one hand and reduce the cost of production at another hand and thus, increase the capacity of farmers and agricultural planners to allocate resources effectively and reduce risks. There is a need for "climate facilitators", who can bridge the gap between science and field application, assisting communities and planners to understand the implications of results for their immediate planning decisions. The imperative of climate change requires increased capacity of farmers to make both, short and long term planning decisions and technology choices.

Existing practice

In general, farmers use their common experience for agricultural planning and more governed by market price will selection of crops and cropping system. However, some farmers are also depended on the locally available traditional wisdom/forecast made by local Bhopas, etc. for the same. But, there is no system or knowledge base for intermittent decision-making regarding various activities in crop production.

Technological intervention

The weather forecast issued by IMD and its implication to agricultural activities is managed to display on black board constructed in the NICRA villages for the purpose. The location and number of the black board has been decided by organizing a meeting of farmers in the village. The village level climate facilitators were prepared to communicate the analyzed information among the farmers. The value added agro advisory service is helpful in improving the climate resilience in agriculture.

Climate facillator interacting with farmers

Director, CRIDA, Hyderabad interacting with project team

Impact

The village level climate fascinators become a key person in the village for disseminating the information right from movement of monsoon to district level forecast of climate scientific date. Because, farmers use this information for crop planning and manage the time of different agricultural operations according to district level climatic scientific data forecast. Three years study results revealed that maize was best performed due to adoption of timely release weather forecasting timely and gave higher yield 2445 kg ha^{-1} with higher net return (Rs.26838) and B: C ratio (3.04) as compared to traditional practice (1971 kg ha^{-1}, Rs.21700 and 2.61), respectively.

In case of sorghum, three years study results revealed that sorghum was best performed due to adoption of timely release weather forecasting timely and gave 24.76 per cent higher yield as compared to traditional practice (1337 kg ha^{-1}). Groundnut was best performed due to adoption of timely release weather forecasting timely and gave 17.59 per cent higher pod yield as compared to traditional practice (1063 kg ha^{-1}). In case of oilseed, three years study results revealed that sesame was best performed due to adoption of timely release weather forecasting timely and gave 40.70 per cent higher yield as compared to traditional practice (113 kg ha^{-1}).

Lesson learned

- Farmers at-least get some information about weather data instead of depending on traditional information.
- Micro level forecast is required for agricultural planning.

SUCCESS STORY-6

6. Improved animal intervention for climate resilience

Animal is an integral component in the farming system of this region. The climate resilience in farming system can be effectively achieved by adopting improved animal suitable for this region.

Existing practice

Farmer use to keep animal as an integral important component for drought proofing and climate resilience in farming system in this region. But the productivity of local breeds of buffaloes is very low. Therefore, there is a need to test the suitability of improved breed of animals in the region to improve the climate resilience in the farming system.

Technological intervention

Improved Murrah breed of buffalo was introduced in the village at three farmers and one improved breed of Murrah he-buffalo was also introduced to change the present breed of buffalo in the NICRA village.

Azolla- A perfect supplemental feed

Improved breed of buffalo Teja Ram Kumawat

HVC, MPUAT, Udaipur interacting with farmers

Impact

The milk yields of these buffalo were 12 to 18 liter per day, which was 150 per cent higher over their local breed. The insemination success rate of the introduced he-buffalo was 80 per cent and he could be used for 216 buffalo in the first year and 298 in the second year. Thus, the murrah breed of buffalo is suitable for this region and supports the existing farming system. The he-buffalo is found to be most effective to change the breed of present buffaloes in the village. This activity will improve the climate resilience of farming system.

Lesson learned

Participatory approach should be used for selection of animal and its procurement.

SUCCESS STORY-7

7. Bio gas plants intervention to improve the environmental condition

It is a proven technology for proper decomposition of Farm yard manure and save the fuel wood for cooking at village level. This is also helpful in lower down the carbon emission in the environment. The adoption of this technology is very low as farmers are not aware having some prejudice thoughts.

Existing practice

The cow dung is heaped in a pit of irregular size with animal feed waste, which results in improper decomposition of farm yard manure and when applied in the field create more initial weed infestation and required more nitrogen to compensate the same. Further, it also invites the problem of termites.

Technological intervention

The concept of "seeing is believing" has been adopted to popularize the proven technology in the NICRA village and a tour of farmer's group has been arranged to a village having bio gas plants. Farmers have interacted with them and came forward to construct the same. In the first phase, eight farmers have been selected to construct bio gas plant having capacity ranging from 2 to 3 CUM according to animal population for improving the quality of manure and to lower down the carbon emission in NICRA villages.

Field day on Bio gas

Technical expert visited the Biogas plant

Impact

Farmers had appreciated the cooking facility with bio gas plants and use the bio gas lamp for lighting also. Further, they have appreciated the outcome i.e. quality of slurry which was much superior to their local system of composting. Even farmers have decided to drill this bio gas slurry instead of spreading as common practice. There was almost no initial weed infestation when they use of bio gas slurry, which is a most labour intensive crucial operation.

Mostly farmers are using the bio gas for the prepare breakfast and thus, there is 900 to 1100 kg wood saving per annum. Bio gas slurry was used in the maize cultivation and there is 70 per cent weed control and 17 per cent yield increase as compared to farm yard manure.

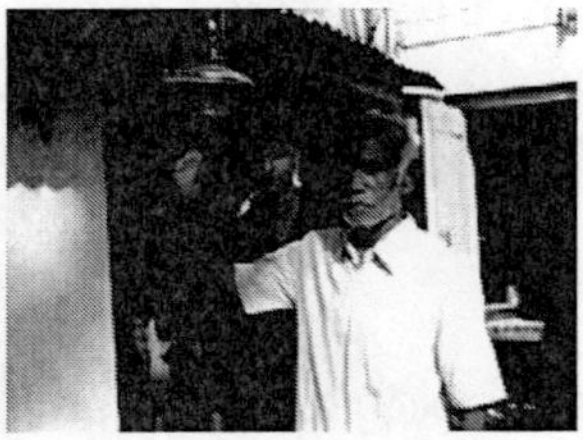

Utilization of bio gas for cooking food

Bio gas plant at farmers' field

Bio gas for using the lighting the lamp

Lesson learned

- Carbon emission can be decrease by adopting bio gas plants.
- Live demonstration is a good tool for adoption of technology.

SUCCESS STORY-8

8. Custom Hiring Centre for improved farm implements

The timeliness and precision in farm operations are very crucial to achieve the desired productivity levels at farmer's field. The continuous dwindling of animal and human power resource in agriculture demands for more adoption of mechanization as compared to traditional agriculture. Further, there are specific research recommendations which require a particular implement like chiesel plough; inter-cropping seed drill, raise based seed drill and rota-til-drill etc. and some implements are required for improving timeliness in farm operation like wheel hoe, power weeder, battery operated sprayers, etc. Adoption of farm implements is a cost intensive task and they prefer to use it before investing their hard money to purchase it.

Existing practice

Popularization of agricultural implements in the region is undertaken by Govt. of Rajasthan through Department of Agriculture by providing subsidies to the farmers for procurement of implements. However, there is no provision of Custom hiring centre, where the farmer can test the implement before investing his hard money. On the other ,the commercial implements are available through private entrepreneur to the farmers but these implements are in general multi-tasking and very poorly maintained which affects the precision in farm operations. These entrepreneurs further hesitate to purchase improved and operation specific implements like planter, rota-til-drill-etc.

Technological intervention

A custom hiring centre was established in one of the NICRA village (Kochriya) to improve the timeliness and precision in different farm operations with a participatory mode. The improved implements viz.; Reversible Disc plough, M B Plough, Chiesel plough, cultivator, Rota-till-drill, Inter-cropping seed drill , Multi-crop seed drill, raise based seed drill, Two row seed drill, Hand seed drill, Arjia wheel hoe, Single row power weeder (Four stroke), Single row power weeder (Two stroke), Three row power weeder, Battery operated power sprayer, duster and Chaff Cutter (Electric). A committee of farmers has been constituted to operate the same.

Reversible disc plough for primary tillage

Director,CRIDA,Hyderabad interacting with farmers on Custom hiring centre

Two row seed drill

Impact

Farmers have appreciated the first time introduction of chiesel plough for breaking the plough pan to increase infiltration and found very effective. Similarly, tractor drawn multi, crop seed drill for sowing of groundnut was also a new introduction in the region, else they here dependent on the bullock power for the same. In the same way, farmers have appreciated the introduction of power weeder for inter-cultural operation and commented that during current agriculture scenario, it is an urgent need looking to shortage of man power as well as bullock power for improving timeliness of operation. The battery operated power sprayer has reduced the drudgery in the operation. Duster was also rarely available in the village and farmer used to adopt crude method for dusting.

The reversible disc plough was a new introduction in the area to minimize the formation of dead furrow. Sowing of recommended inter-cropping of maize+black gram can only be managed by bullock power only and can not be sown by traditional tractor drawn seed drill, further reduction in availability of these hinders the adoption of recommended inter-cropping system. Therefore, tractor drawn inter-cropping seed drill developed at this station has been introduced in the area. Similarly, for sowing of groundnut crop farmers are dependent on bullock power only; therefore, tractor drawn multi-crop seed drill has been introduced in the area. The farmers have appreciated the uniformity plant population obtained in each crop. The Arjia wheel hoe developed at this station has been used widely by the farmers in an area. Similarly, introduction of power weeder of different type has been a miracle for the farmers, as it was an acute need due to low availability of time and labour during inter-culture operation. The battery operated power sprayer was also new for the farmers of this region and they have appreciated the low fatigue in spraying of pesticides and weedicides.

Battery operated sprayer

Power weeder- 70cc engine

Power weeder- 5HP engine sprayer

Power weeder-31cc engine

Compartment of seed block of Inter-cropping seed drill

The custom hiring centre was used by 195 and 186 farmers in 72.1 and 82.9 ha area during the year 2012-13 and 2013-14, respectively. Overall, the farmers of the village have appreciated the effectiveness of the custom hiring centre for creating awareness about improved implements and improving the availability of implements thereby improving the precision and timeliness in farm operation. The farmers from the nearby village also used to come there to see the new implements and they have also demanded for such scheme from their political leaders.

Lesson learned

- An effective source for introduction of new implements and provide an opportunity to the farmer to use the implement before investing his hard money
- Effective for improving timeliness and precision in farm operations
- Management absolutely at farmer's level seems not feasible
- Not good for traditional popular implements

- Social management is an integral part for successful operation of the CHC.
- The model "Custom Hiring Centre "has been appreciated by farmers and it has demand at Block Level at least.

SUCCESS STORY-9

9. Silvipastoral model for improving productivity of grasslands

Farmers face acute shortage of fodder in drought years because of poor carrying capacity of their pasture lands. The pre-dominant reason for poor carrying capacity of these pasture lands are improper management (open grazing without any system) and overgrazing the grass, non-adoption of rain water conservation practices, and poor quality grass.

Existing practice

Sh.Kishan s/o Hema Jat, Bhagatpura, panchayat -Lapshiya,P.S. Railmagra, Distt. Rajsamand-Rajasthan had 1.2 ha land kept for pasture land as per common practice to keep some fellow land as pasture depending upon the animal population. But, carrying capacities of this land are very low due to adoption of local quality grass seed and no measures have been adopted for rain water conservation.

Technology intervention

Rain water conservation practice (contour trenching at 8 m horizontal interval) and forage grass (*Cenchrus setigerus -CAZRI-76*) to develop his unproductive pasture land. Due to intervention, an increase of 73.9 per cent in the productivity of dry grass yield as compared to farmer's practice with a net return of Rs. 13950/ per hectare. The quality of the produce is also improved with regard to nutrition and palatability.

A view of pasture land before development

Developed pasture land

Impact

Farmers of the village have been so convenience about the technology that they have come forward to form an SHG 'Shrinath Help Group" for development of their earmarked community pasture land and committed to maintain the same. Improved grasses (*Cenchrus setigerus and Stylo hamata*) with rain water management practice performed better at farmers field to the tune of 213% with higher B:C ratio as compared to local grass (2200 kg ha^{-1}).

Farmer-farmer interaction on developed pasture

Field day on pasture land

Lesson learned

- Successful demonstrations have great impact.

SUCCESS STORY-10

10. Community seed bank for improving seed replacement rate

Community seed bank in the village are essential tool for adoption of dynamic crop planning based on onset of monsoon. For example, if monsoon is delayed, it is recommend to adopt short duration variety of maize or go for fodder, but availability of such improved seeds at the last moment hinders the adoption of such recommendations. In such conditions, the community seed bank can play a crucial role in improving the availability of desired seed. The seed replacement rate can be effectively improved with the help Community seed bank. Further to saturate a particular drought is a recurring phenomenon and under climatic change scenario, the intensity of extreme events is more severe. The community seed bank is also helpful to saturate a particular variety of crop in the village at reasonable cost. This concept also reduces the cost of cultivation.

Existing practice

Trials and demonstrations are conducted to select an appropriate crop and its variety under climate change scenario in this region and accordingly farmers are suggested to procure the recommended variety from the market and adopt it. The dependency in timely availability of a particular variety in general defeats the basic idea of selection of that variety. Further, the cost of the seed is much higher as compared to producing the same at village level.

Technological intervention

A group of farmers having piece of lands in a particular area selected for seed purpose have been formulated under the NICRA village Kochariya, Block–Suwana, District–Bhilwara. At initial level, farmers have adopted the new variety of groundnut for development of a seed bank looking the high cost and lower availability of improved seed in the village. The groundnut variety TG37A have been selected for seed production in the village.

Scientist farmer interaction on seed bank

Bumper harvest of Groundnut TG-37A

Impact

Farmers of the village gain confidence that they can also produce the seed of the desired composite variety at village level and get the advantage in yield and price. It is also helpful in reducing the cost of cultivation and improves the adoption of improved variety of different crops for climate resilient agriculture.They have formulated a group of 45 farmers for sowing of foundation seed of improved groundnut variety TG37A in an area of 12 ha and produce 125 q of seed. Out of total production, farmers have exchanged the seed around 10 q and 12 q seed has been kept with farmers for self-sufficiency and deposited 6.3 q of seed in the village community seed bank. Similarly, a cluster of farmers have been promoted to develop a seed village for maize variety PM-3 in an area of 20 ha. Similarly, a cluster of farmers have been promoted to develop a seed village for maize variety PM-3 in an area of 20 ha.

Lesson learned

- Community seed bank and village play a crucial role for improving the seed replacement rate and climate resilience in agriculture
- Community mobilization can be achieved for mutual benefit
- Cost of production can be reduce by producing seed at village level
- This activity promotes communal harmony among the farmers

SUCCESS STORY-11

11. Community grassland developments through farmer's participatory approach

There is provision of keeping sufficient earmarked pasture land in almost each village to support the livestock population of that village by the state government of Rajasthan. Inspite of this, farmers are facing the acute shortage of fodder in drought years because of poor carrying capacity of these earmarked pasture lands. The pre-dominant reason for poor carrying capacity of these pasture lands are improper management (open grazing without any system), non-adoption of rain water conservation practices, poor quality seed and overgrazing the grass which does not provide opportunity for seed multiplication and the most prominent is lack of ownership feeling of these lands by the community.

Old status of community pasture land

DR, MPUAT, Udaipur interacting with farmer on community pasture land

Formation of ditch cum bund fencing with cactus through farmers participation

Existing practice

The earmarked pastures lands are being put to grazing by animals of the village in a random manner without any system of grazing like rotational grazing or limited grazing or managing the time of grazing. This leads to uncontrolled grazing which results poor carrying capacity of such lands. Further, being a common property resource no body takes care to adopt improved technologies for the same.

Technological intervention

To address this problem, first step was taken to show the potential carrying capacity of these lands to the farmers as mentioned above the success stories and conducting number of meeting, focused group discussions to motivate the farmers about development of community earmarked pasture land. The technological interventions include contour trenches at 8 m horizontal for rainwater conservation and cross cultivation with cultivator were carried out to put grass seed of improved variety Cazari-76 in lines at 45 cm apart. Farmers have selected the Neem (*Azadircta indica*), desi babool (*Acacia species*) and Ingadalsis/keekar (*Piethocolobium dulce*) at a spacing of 5 m in the contour trenches. Farmers have used the cut and carry method of harvesting the grass.This year, the villagers have taken 12.8 ha of land put all their efforts to develop these earmarked pasture land out of their total 44 ha.

They have come one step ahead of their commitment by creating a water harvesting structure (Nadi) to rain water for animal drinking and groundwater recharge. As per their commitment, they are managing watering to the plants and keeping watch and ward of the developed area. They have further agreed to put all the earmarked pasture lands in step-by-step in four years to adopt for rotational grazing system. The modalities of community works have been adopted very differently for development. The work has been carried out by the farmers through self help group in technical collaboration with scientist of this station and payment (project contribution) has been made to the self help group. This system gives farmers a satisfaction of ownership of work being carried out in their earmarked pasture land and every farmer of the village was seen very vigilant about the work being carried out. Some of them are appreciating and boosting the owners/drivers by providing them timely tea and breakfast as if they are working at their own fields. The success of development of earmarked pasture land by the farming community for the farming community is a very rare example and if adopted on large scale.

Contour trenching at 8.0m HI for rainwater conservation

Line sowing of grass seed through farmers participation

Plantation of pasture forage tree

Impact

The development of earmarked pasture land was very encouraging looking to the growth of the grass and tree species in the area which had not been seen earlier in the village. The farmers have maintained the social-cum-bio fencing and keep it free from grazing of animals. The harvesting of the grass was done by cut and carry method and managed by the farmers. This year a grass yield of qt per ha was recorded as compare to no yield in previous years.

Director, CRIDA, Hyderabad interacting with scientist & farmers on community pasture land

A view of developed community silvi pasture land

HVC, MPUAT, Udaipur interacting with scientist and farmers

Lesson learned

- Community mobilization can be a good tool to deal to improve the productivity of earmarked pasture lands.
- It is the best bets to mitigate the effect of climate change on the animals by making village self-sufficient in fodder production.

SUCCESS STORY-12

12. Development of community rainwater harvesting structure through farmer's participation

The climate change adversely affects the availability in rainwater to human beings and animals. Therefore, small new rainwater harvesting structures are being created but farmers are not taking care of existing community rainwater harvesting structures in the village. The rainwater harvesting, if adopted at community level, has more impact as compare to small farm ponds constructed with respect to climate resilient agriculture. These structures improve the ground water level and surface water is available for drinking of animals nearby earmarked pasture land.

Existing practice

There is no surface water source available for animal drinking nearby earmarked pasture land and availability of groundwater is also very scarce for taking rabi crops. A seasonal Nallah is flowing within the earmarked pasture land but there is no any provision to harvest the rainwater.

Technological intervention

Community mobilization has been achieved to construct an earthen embankment for harvesting the rainwater to improve the groundwater recharge and availability of drinking water for animals. Farmers of the NICRA village Bhagatpura, Panchayat–Lapsiya, Tehsil –Railmagra, District Rajsamand were motivated to develop rainwater harvesting structure-Nadi for harvesting the rainwater. The length of the earthen embankment is 285m with an average dimension: top width-3.07m (2.3–3.7m), height–1.44m (0.3–2.1m), bottom width-5.95 m (2.9m-7.9m) Approx. Capacity–20000cum

Director Research, MPUAT, Udaipur visited the community NADI

Old status of community NADI

A view of community NADI

Impact

The intervention has improved the groundwater recharge in nearby area for climate resilience in agriculture and the water availability for animal drinking has also been improved. As an impact of the aforesaid intervention, farmers of the nearby village also approach for development of such structure.

Lesson learned

The community mobilization activity cannot be a planned activity in any programme or project as when it become a mandatory activity it rather difficult to implement the same at village level. A source of fund should be created at centre/state level for promotion of community mobilization activity.

Conclusion

The Indian Meteorological Department, Government of India is regularly releasing district level forecast of climatic parameters like rainfall, minimum and maximum temperature, minimum and maximum relative humidity, cloud cover (Octa) and wind speed and direction for consecutive five days to reduce the uncertainties for agricultural planning. The information so generated is useful to improve early warning systems for drought, flood, and pest and disease incidence and thus, increase the capacity of farmers and agricultural planners to allocate resources effectively and reduce risks. There is a need for "translators" of climate information, who can bridge the gap between science and field application, assisting communities and planners to understand the implications of results for their immediate planning decisions. Enhancing communication between producers and users of climate science is also clearly a requirement. Institutions to facilitate this exchange can be existing communications and information dissemination networks through climate facilitators. Finally platforms for collaborative action and information sharing, such as the we Adapt platform, which unites modellers, practitioners and donors, can enhance the development and use of climate science information for agricultural decision-making .The imperative of climate change requires increased capacity of farmers to make both, short and long-term planning decisions and technology choices.

❒❒❒

Chapter 12

Improvement of Fruit Crops Suitable for Arid and Semi-arid Regions

Dhurendra Singh

Division of Crop Improvement, ICAR-Central Institute for Arid Horticulture, Bikaner-334006

Introduction

Indian horticulture sector contributes about 38% to the agriculture Gross Value Added (GVA) making very significant contribution to the Indian economy. Apart from ensuring nutritional security of the nation, it provides alternate rural employment opportunities, diversification in farm activities, and enhanced income to farmers. India is currently producing about 283.5 million tonnes of horticulture produce which has surpassed the food grain production, that too from much less area (24.2 m ha. for horticulture against 125 M. ha. for food grains). Productivity of horticulture crops is much higher compared to productivity of food grains (11.96 tonnes/ha against 2 tonnes/ha.). India has emerged as world leader in the production of a variety of fruits like mango and banana, and is the second largest producer of fruits and vegetables.

The hot arid regions are spread over about 31.7 million ha area mainly in the States of Rajasthan, Gujarat, Andhra Pradesh, Punjab and Haryana, which inhabit on an average 61 persons per square km making up a population of nearly 20 million people. The Indian arid zone is characterized by high temperature and low and variable precipitation which limit the scope for high horticultural productivity.

The arid horticultural crops have developed and/or modified their organs to perform certain vital physiological functions, such as strong deep root system (ber, bael, aonla, wood apple, jamun, etc.), synchronize their flowering and fruit development with the season of moisture availability (ker, lasora, aonla, pilu, etc.) and other xerophytic characters i.e. leaf shedding in summer (ber), scanty foliage (ker), spiny cladode (cactus pear), mucilaginous sap in plant part (ker, gonda, pilu, bael, etc.), sunken stomata and fur/hairiness and waxy coating on the leaf surface (phalsa, ber, gonad, fig, etc.) thorny nature, and selective or reduced absorption of cation (Na^+) and anions (Cl^-, $SO4^-$) for survival under adverse arid conditions.

Epidemiological studies indicate that a diet rich in fruits and vegetables imparts health benefits. For this reason, health organizations around the globe recommend the intake of several servings of fruits and vegetables each day.

Word wide attempts on development of value added products as nutraceuticals through intensive research and development efforts are concentrated in developed countries and translated into commercial ventures. Global market for nutraceuticals is projected to surpass US$187 billion in sales by 2010. In India, the market for functional foods is targeted to cross Rs. 17,000 crore in the next four years. Nutraceuticals are based on bioactive plant constituents found abundantly in fruits, vegetables, other horticultural crops including generally regarded as safe (GRAS) plants. However, the availability of these bioactives in plants is genotypical and dictates by multitudes of environmental factors. A wealth of scientific literature from numerous epidemiological and case controlled studies have identified the potential relationships between bioactive compounds (or "functional" components) and their protective effects against hypertension, cardiovascular disease, cancer, and other adverse health conditions.

Table 1: Value added products from various arid fruit crops:

Fruit	Products
Aonla	Candy, Jam, Syrup, Sweets, Pickle, Chutney, Squash, Ayurvedic medicines, such as chavanprsh, Trifla, Amla ki Rasayan and Powder.
Ber	Candy, Dried ber, Squash.
Custard apple	Juice, Squash, Wine, etc,.
Pomegranate	Juice, Squash, Dried seeds as 'Anardana'.
Wood apple	Jelly, Chutney, Squash.
Guava	Mainly eaten fresh, Jam, Jelly, Paste, etc,.
Tamarind	Juice, Pulp.
Date palm	Pulp is suitable for making 'Chhuhara', Syrup.

Fig	Dried fig.
Jamun	Jelly, Squash, Vinegar, Wine.
Karonda	Pickle, Chutney, Candy, Syrup and Jelly.
Bael	Candy, Squash, Jam Powder.
Mahua	Alcohol, Bakery products, Vinegar and Syrup after fermentation may be used as cattle feed. Pulp dry husk for making alcohol.
Ker	Pickle.
Lasora	Pickle.

Selection of horticultural crops and their varieties

While selection of fruit crops for dry land horticulture, one of the basic requirements is that those crops who complete their vegetative growth and reproductive phase during the period of maximum moisture availability. The fruit, such as ber, guava, pomegranate, custard apple, aonla and sour lime, conform to this prerequisite. Among vegetable crops, the hardiest types belong to the cucurbitaceous and solanaceous groups. Cowpea, cluster bean and okra are also very drought hardy. In these crops also, there is need to select the most drought hardy cultivars for rainfed production. Varieties were also identified in fruits, vegetables, flowers and medicinal and aromatic plants through researches conducted in different parts of the country. The important crops and cultivars which can be grown successfully in hot arid region are presented below.

Improvement in fruit crops suitable for arid and semi-arid zone

Goma Kirti : Goma Kirti is high yielding Late maturing variety

Thar Sevika : Developed by the hybridization from a cross Seb x Katha, Thar Sevika is an early maturing variety. Average fruit yield is 30-32 kg/tree.

Thar Bhubhraj : A selection from local material of Bhusavar area of Bharatpur district of Rajasthan having an average yield potential of 30-36 kg/tree. The fruits are sweet and TSS content 22-23%.

Thar Malti : Fruit yield: 65-70 kg/plant; Fruit weight: 20.32g, TSS: 18.6 °Brix, Acidity: 0.46 %, Vitamin C: 126.7 mg/100g, Total flavonoids: 2.49 mg/g FW, Total polyphenols: 2.93 mg/g FW, Total antioxidant activity (CUPRAC): 13.1μM TE/g. Tolerant to low temperatures (frost; subzero -2 °C) as plant growth.

Goma Yashi : Produce very good quality fruit with weight of 1.25 kg per fruit.Fruit are ovate in shape, greenish yellow. Flesh colour is Straw colour. Fruit shell weight is 180g. Number of locules is cross section/fruit is 17.

Thar Divya : Plant height (vigorous in growth pattern), semi-spreading growth habit, compact canopy, very less spines, fruit ripens in 270 days from fruit set. Comparatively earliest in maturity (second fortnight of February), precocious bearer and highly suitable for growing under dryland/rainfed hot semi-arid ecosystem (under 8m x 6m distance). Drought tolerant, luxuriant growth and higher fruit yield under less precipitation and high temperature. Attractive dark yellow colour of pulp of complete ripened fruit. No major diseases and pests were noticed under field conditions.

Thar Neelkanth : This variety was developed at ICAR-CIAH regional station CHES, Godhra, Gujarat during the year 2016. It is superior genotype with desirable characters like compact growth, semi-spreading, medium height (5.0 m) with few spines, and better yield with quality fruits having pleasant flavour with attractive colour of pulp (dark yellowish) and uniform fruit size under rainfed semi-arid environment.

Aonla : Goma Aishwarya A selection from plus tree, Goma Aishwarya is an early, drought tolerant. The average yield potential is 105 kg/tree.

Pomegranate : Goma Khatta: The variety is developed for Anardana purpose. Yield potential is 6.59 kg/plant and anardana yield is 1.18 kg/plant. Seeds hardness is medium. Fruit having 46.7% of Juice and TSS is 14.5^{0}Brix. Acidity is 7.3%.

Jamun : The Jamun (Syzygium cuminii Skeels) is a nutritious fruit with a variety of uses. The fruit is good source of iron, sugars, minerals, protein and carbohydrate, etc. Fully ripe fruits are eaten fresh and can be processed into beverages like jelly, jam, squash, wine, vinegar & pickles. Seeds contain alkaloid jambosin and glycoside, which reduce/ stop the diastatic conversion of starch in to sugars. Fruits are used as an effective medicine against diabetes, heart and liver trouble.

Goma Priyanka : Goma Priyanka was released at CHES, Godhra in the year 2010. The selected genotype was propagated through patch budding and tested under field conditions for 8 years (2001-2009). The selection out performed in respect of flowering pattern, fruiting and fruit quality attributes. The tree height and girth was recorded 5.25 m and 64.66 cm respectively. North-South and East-West spread was found to be 5.00m and 4.76 m respectively.

Thar Kranti : Developed during the year 2016. It is dwarf. Fruit ripens in 75 days from fruit set. Fruit yield: 65.00 kg per plant. Fruit shape: Oblong, Fruit colour: Deep purple, Flesh colour: Dirty white, Fruit weight: 20.10g, Pulp weight: 17.20 g, Pulp per cent: 85.57%, TSS: 17.10 ^{0}Brix, Acidity: 0.40%, Total sugar: 12.50%, Reducing Sugar: 6.20%, Vitamin C: 48.45 mg/100g.

Other promising genotypes of jamun

GJ-8 : It was collected from Ode village of Anand district of Gujarat. Peak period of flowering is in the month of March. It ripens in the second week of June and recorded 17.00 g average fruit weight, 81.82 per cent, pulp, 14.20 per cent, TSS, 0.39 per cent, acidity, per cent, 11.35 per cent, total sugar and 45.10 mg/100 g vitamin C.

GJ-40 : It was also collected from Ode village of Anand district of Gujarat. Peak period of flowering is in the month of March. It ripens in the fourth week of June and recorded average fruit weight 16.00 g, 82.00 % pulp, 14.00 % TSS, 0.39 % acidity, 11.00 % total sugar and 46.10 mg/100g vitamin C.

Tamarind

Tamarind (Tamarindus indica L.) a member of sub-family caesalpiniaceae of family leguminoceae, is an under-utilized fruit and widely distributed throughout tropic and sub-tropics as stray plantation or avenue. Owing to its varied use in household and export market, it has emerged as a potential fruit crop. At CHES, Godhra, total 24 genotypes were evaluated (2001-2009) for their growth, flowering and fruiting attributes and Tamarind-13 (Goma Prateek) was screened as promising one. Salient characters of Goma Prateek is as under:

Goma Prateek

Goma Prateek was released at CHES, Godhra in the year 2010. It is early, spreading growth habit, regular bearer, semi-dwarf and starts flowering in 4th year. Peak period of flowering is first week of June and ripens in the first week of March and recorded 26.70 g average pod weight, 16.70 cm pod length, 50.50 per cent pulp, 14.06% acidity,

71.00°Brix TSS, 55.81% total sugar, 27.27% reducing sugar, 17.53 mg/ 100g vitamin C, 3.29g/100g protein, 177.70 mg/100g calcium, 44.17 mg/100g magnesium, 70.83 mg/100g phosphorus. Fruit shape is slightly curved with reddish pulp, suitable for processing.

DTS-1 : It is a selection identified at University of agricultural Sciences, Dharwad. The pods are straight having semi-curved shape, 23.6 cm length, 3cm width, 19.5 gm weight, 51.00% pulp and 13.60% acidity. It is a late variety and takes 310 days from fruit set to maturity.

DTS-2 : This is a selection made at University of Agricultural Sciences, Dharwad. The pods are straight having semi-curved shape, 17.60 cm length, 2.60 cm width, 18.00 gm weight, 53.00% pulp and 12.20% acidity. It is an early variety and the pods mature in 280 days after fruit set.

Chiraunji

Chiraunji (Buchanania lanzan Spreng.) is excellent tree of agro forestry and social forestry. In the wasteland development and dryland horticulture, it assumes great significance due to its multifarious uses and capacity to withstand adverse climatic conditions. The flesh of ripen fruit is very palatable and is largely eaten raw or roasted and the oily kernels are the most important part and are used in preparation of puddings. Very good juice may be prepared from the pulp of chiraunji fruits. The kernel is highly nutritious and rich in protein (25.0-30.0%) and yields sweet oil, which can be used to substitute olive and almond oil. Kernel contains 33.50% oil, 1.90% of which is unsaponifiable. The saponifiable part contained 20.00% of linoleic acid. An ointment made out of the kernels is used to cure itch of the skin and to remove blemishes from the face.

Thar Priya

Plant height (3.43 m), semi-spreading growth habit, thick trunk, dense foliage and dropping branches, umbrella shape, fruit ripens in 50-65 days from fruit set. It is comparatively dwarf, precocious bearer and suitable for high density planting. No major diseases and pests were noticed under field conditions. It is suitable for table and processing purpose due to high pulp content, TSS, sugar and kernel content. It recorded 23.90 °Brix TSS, 1.24% acidity, 13.06% total sugar, 6.67% reducing sugar, 48.70 mg/100g vitamin C and 31.36% kernel protein.

Mahua

Survey of Mahua was made in Gujarat. It is one of the important crops for tribal farmers of Gujarat. Detailed description of 4 promising genotypes are as under:

Mahua-10

It was collected from Vejalpur village of Panchmahal district, Gujarat. Peak period of flowering is in the second week of March. Flowers recorded 2.29g weight, 65.00% juice and 26.37% TSS. It ripens in second week of May and recorded 30.50 fruit weight, 13.50g seed weight and 11.00g kernel.

Mahua-14

It was collected from Otala village (Baria Road) of Panchmahal district, Gujarat. Peak period of flowering is in the second week of March. Flowers recorded 2.24g weight, 66.00% juice and 25.00% TSS.

Mahua-35

It was collected from Reechawani village (Baria Road) of Dahod district. Peak period of flowering is in the second week of March. Flowers recorded 2.13g weight, 64.12% juice and 24.90% TSS.

Mahua-63

It was collected from Reechawani village (Baria Road) of Panchmahal district, Gujarat. Peak period of flowering is taken place in the 4th week of March. Flowers recorded 2.50g weight, 68.00% juice and 24.37% TSS.

Thar Madhu

Plant height (4.40 m), semi-spreading growth habit, thick trunk with dense foliage. No major diseases were noticed under field conditions. Plant growth and fruit yield was almost normal during the years received less precipitation.

Trait specific

Yield : 20.10 kg per plant during 10th year of orchard life under rainfed conditions of hot semi-arid ecosystem. Yield of dry flower: 6.30 kg per plant. Fruit weight: 29.0g

Maturity : Fruit ripens in 90-105 days from fruit set (Ripe fruits shed from trees during May-July).

Uses : Mahua flowers used in preparation of distilled liquors. It is used for preparation of cake, vinegar, jam, syrup and honey. Flowers and spent flowers (after fermentation) are utilized as feed for livestock. Seeds processed for oil are used to manufacture vanaspati ghee, soap, greases and cosmetics and also used in bakery products, as manure and has insecticidal properties.

Khirni

Khirni or Rayan botanically known as (Manilkara hexandra) is an economically multipurpose tree of the family sapotaceae. Thirty genotypes were selected & evaluated their fruiting and fruit quality attributes. Detailed description of promising genotypes are as under:

GK-1

It was collected from Kharsalia village of Panchamahal district, Gujarat. The peak period of flowering was recorded in the month of November. It ripens in the first week of March and recorded 5.66g fruit weight, 4.93g pulp weight, 21.00% TSS, 16.20mg/100 vitamin C and 4.30g/100g carotene.

GK-10

It was collected from Parwadi village of Panchamahal district, Gujarat. The peak period of flowering was recorded in the month of December. It ripens in third week of May and recorded 6.64g fruit weight, 5.71g pulp weight, 26.40% TSS, 24.25mg/100 vitamin C and 6.6.06g/100g carotene.

Thar Rituraj

Plant height (4.36 m), semi-spreading growth habit, thick trunk, evergreen & dense foliage umbrella shape. It is semi-dwarf, precocious bearer, flowering start in 4th year, fruit ripens in 120-125 days from fruit set. It is comparatively dwarf, precocious bearer and suitable for high density planting.

Antioxidant value of some superior Khirnee genotypes/accessions

The maximum antioxidant value was observed in accession CISH K-3 (213.85 µg/ml AEAC unit) using Ferric reducing antioxidant potential, while CISH K-1 showed minimum value (78.05 µg/ml AEAC unit). The total carotenoids was recorded maximum in CISH K-2 (85.9 µg/100ml) and minimum in 9.8 µg/100 ml in CISH K-3.

Karonda

Thar Kamal : The variety is developed through selection from existing germplasm at CHES-CIAH, Godhra. The selected genotype was tested under field conditions for 9 years (2005-2014). Plant height (1.77 m), semi-spreading growth habit, thick trunk, evergreen, dense foliage and drooping branches.

Mulberry

Thar Lohit : This variety was developed by ICAR-CIAH, Bikaner during the year 2016 with a salient features of plant height (3.07 m), spreading growth habit, heterophyllous leaves cordate in shape and with fine serrated margin, leaf pubescence absent, leaf arrangement alternate and red colour, pendulous shape fruits. planting distance: 6x6 m, tree population : 275/ha.

Thar Harit : Fruit maturity duration: 38-42 days, Fruit yield: 32.6 kg per plant, Fruit weight: 3.76g, TSS: 31.5 °Brix, Acidity: 0.42%, Vitamin C: 23.7mg/100g, Total flavonoids: 0.72 mg/g FW, Total polyphenols: 0.84 mg/g FW and Total antioxidant activity (CUPRAC): 4.30µM TE/g, Tolerant to low (frost) and high temperatures (from subzero (-2 ^{0}C) to 48 ^{0}C).

Thar Pragati : The variety is developed through selection from existing germplasm at CHES-CIAH, Godhra. Fruit ripens in 60 days from fruit set. Fruit ripening start from 2nd week of April. Yield: 3-4kg per plant, Fruit weight: 2.1g, pulp weight: 1.9g, Pulp: 90.4%, fruit shape: Round, Fruit Colur: Deep purple, TSS: 20.12 ^{0}Brix, Acidity: 3.12 %, Total sugar: 12.42 %, Reducing Sugar: 8.10%, Vitamin C: 19.12 mg/100g.Plant height: 2.20 m, spreading growth habit, thick stem, dense foliage and drooping branches.

References

1. Singh Sanjay and Singh, A K. 2005a. Genetic diversity in mahua (Bassia latifolia) under semi-arid ecosystem of Gujarat. Indian J. Agril. Sci., 75: 519-23.
2. Singh Sanjay and Singh, A K. 2005b. Genetic variability in jamun (syzygium cumini skeels) from Gujarat. Progressive Hort.37: 44-48.

3. Singh Sanjay and Singh, A K. 2005c. Studies on genetic diversity in Mahua(Bassia latifolia) under semi-arid ecosystem of Gujarat. Indian J. Agril. Sic. 75: 519-523.

4. Singh Sanjay, Singh, A K and Apparao, V. V. 2006. Genetic diversity in Chironji (Buchanania lanzan) under semi-arid ecosystem of Gujarat. Indian J. Agril. Sic. 76: 695-698.

5. Singh D, Krishna H, Singh R S, Singh U V and BD Sharma (2016) Mulberry: A promising fruit crop for arid regions CIAH/Tech./ Pub. No 62, ICAR, CIAH, Bikaner, Rajasthan, pp 1-18.

6. Saroj P.L., Singh R S and Krishna H. (2017) ICAR-CIAH-An Overview. Pub.CIAH/Tech./Pub./No.60.

❏❏❏

Chapter 13

Grass–Legume Inter-cropping in Arid and Semi-arid Regions

S.M. Kumawat

Prof. Agronomy, Department of Agronomy, Swami Keshwanand Rajasthan Agricultural University, Bikaner-334 006

Introduction

Arid and semi–arid regions mostly drylands, have been cradles to some of the world's greatest civilizations. These dryland ecosystems support a surprising amount of biodiversity. Desertification, however, is a significant land degradation problem in the arid, semi-arid and dry sub-humid regions of the world. Deterioration of soil and plant cover has adversely affected 70% of the world's drylands as a result of extended droughts as well as mismanagement of range and cultivated lands. The situation is likely to worsen with high population growth rates and accompanying landuse conflicts.

Intercropping has been important in the US and other countries and continues to be an important practice in developing nations. In traditional systems, inter-cropping evolved through many centuries of trial and error. To have persisted, intercropping had to have merit biologically, environmentally, economically, and sociologically. Any agriculture practice must provide advantages over other available options in the eyes of the practitioner to gain acceptance.

Principle : Diversity is nature's design

When early humans replaced hunting and gathering food with growing crops and animals, the landscape changed accordingly. By producing a limited selection of crop plants and animals, humankind has reduced the level of biological diversity over much of the earth to much simpler levels. Annual crop monocultures represent a classic example. In response to this biological simplification, nature has struggled to restore diversity to these landscapes – that is her tendency. Our "war" with nature over the tendency to diversity is what we call "weed control" and "pest management."

Principle : Co-operation is more apparent than competition

There is far more cooperation in nature than competition. Co-operation is typified by mutually beneficial relationships that occur between species within communities. In the Re-designed Forest, ecologist Chris Maser offers a glimpse of the co-operation inherent in a northern temperate forest when he describes a relationship that exists among squirrels, fungi, and trees (Maser,Chris.1990). The squirrels feed on the fungi, then assist in its reproduction by dropping fecal pellets containing viable fungal spores onto the forest floor. There new fungal colonies establish.

Principle : Stability tends to increase with increasing diversity

Once a wild plant and animal community has reached a high level of diversity, it appears to remain stable for many years. When wild communities are in the early stages of development, or when they have lost diversity due to natural catastrophe or human actions, they are prone to major fluctuations, both in type of species present and in their numbers. Disease outbreaks in plants and animals occur more frequently – as do outbreaks of weed, insect, bird or rodent pests. One good example is the grasshopper plagues that follow regional weather shifts.

Pursuing Diversity on the Farm

Some pioneering farmers have been able to utilize nature's principle of diversity to their advantage. Results of their efforts include lower cost of production and higher profits. Among the practices that promote diversity and stability are:

Enterprise diversification : Stability of income and yield are two of the reasons people diversify their crop and livestock systems. Increasing diversity on-farm also reduces costs in pest control and fertilizer needs because these costs can be spread out over several crops or animal enterprises.

Integration : On-farm diversity can be carried to an even higher level by integrating animals with intercrops as harvesters. With each increase in the level of diversity comes to an increase in stability.

Crop Rotation : Moving from simple monoculture to a higher level of diversity begins with viable crop rotations, which break weed and pest lifecycles, and provide complementary fertilization to crops in sequence with each other.

Farmscaping : Diversity can be increased again by providing more habitat for beneficial organisms with borders, windbreaks, and special plantings for natural enemies of pests.

Inter-cropping : Inter-cropping is the growing of two or more crops in proximity to promote interaction between them. Much of this paper focuses on the principles and strategies of inter-cropping field crops, including grass-legume forages under arid and semi-arid regions.

Intercropping Concepts

Most grain-crop mixtures with similar ripening times cannot be machine-harvested to produce a marketable commodity–since few buyers purchase mixed grains. Because of limited harvest options with that type of inter-cropping, farmers are left with the options of hand harvesting, utilizing crops in the field with animals, or harvesting the mixture for on-farm animal feed. However, some inter-cropping schemes allow for staggered harvest dates that keep crop species separated.

When two or more crops are growing together, each must have adequate space to maximize co-operation and minimize competition between the crops. To accomplish this, four things need to be considered: 1) spatial arrangement, 2) density, 3) maturity dates of the crops being grown, and 4) plant architecture.

Spatial Arrangement

There are at least four basic spatial arrangements used in inter-cropping. These are:

- Row inter-cropping : growing two or more crops at the same time with at least one crop planted in rows, example-barley + mustard.
- Strip inter-cropping : growing two or more crops together wide enough to permit separate crop machine production but close enough for the crops to interact, example–maize+soyabean/urd; sorghum+cowpea/guar.

- Mixed inter-cropping : growing two or more crops together in no distinct row arrangement, example–lucerne+Chinesecabage; oats+lucene/berseem.
- Relay inter-cropping : planting a second crop into a standing crop at a time when the standing crop is at its reproductive stage but before harvesting, example–rice+ricebean.

Density

To optimize plant density, the seeding rate of each crop in the mixture is adjusted below its full rate. If full rates of each crop were planted, neither would yield well because of intense overcrowding. By reducing the seeding rates of each, the crops have a chance to yield well within the mixture. The challenge comes in knowing how much to reduce the seeding rates. For example, if you are planning to grow corn and cowpeas and you want mostly peas and only a little corn, it would be easy to achieve this.

Maturity Dates

Planting intercrops that feature staggered maturity dates or development periods takes advantage of variations in peak resource demand for nutrients, water and sunlight. Having one crop mature before its companion crop lessens the competition between the two crops. An aggressive climbing bean may pull down corn or sorghum growing with it and lower the grain yield. Timing the planting of the aggressive bean may fix the problem, if the corn can be harvested before the bean begins to climb. When the corn was mature, the beans or peas had corn stalks to climb on.

Plant Architecture

Plant architecture is commonly used strategically to allow one member of the mix to capture sunlight that would not otherwise be available to the others. Widely spaced corn plants growing above an understory of beans and pumpkins would be a classic example.

Intercrop Productivity

One of the most important reasons to grow two or more crops together is the increase in productivity per unit of land. Researchers have designed a method for assessing intercrop performance as compared to pure stand yields. In research trials, they grow mixtures and pure stands in separate plots. Yields from the pure stands and each separate crop from within the mixture are measured. They then know how much additional land is required in pure stand to equal the amount of yield achieved in the inter-crop. The calculated figure is called the Land Equivalency Ratio (LER). To calculate an LER, the intercrop yields are divided by the pure stand yields for each component crop in the inter-crop. Then, these two figures are added together (inter-crop corn/pure corn) + (inter-crop pea/pure pea) = LER

When an LER measures 1.0, it tells us that the amount of land required for peas and corn grown together is the same as that for peas and corn grown in pure stand (i.e. there was no advantage to inter-cropping over pure stands). LERs above 1.0 show an advantage to inter-cropping while numbers below 1.0 show a disadvantage to inter-cropping. For example an LER of 1.25 tells us that the yield produced in the total inter-crop would have required 25% more land, if planted in pure stands. If the LER was 0.75 then we know the inter-crop yield was only 75% of that of the same amount of land that grew pure stands.

Managing Inter-crops

Many combinations of crops have been grown or experimented with as mixed or relay intercrops. Some of these combinations include cucumbers, beans, celery, and chives in China; and corn, peanut, sorghum and millet, and sorghum + pigeonpea grown together in various parts of the tropics.

Frequently these cropping combinations involve a short and a tall crop both planted at the same time. In many cases, the tall crop is harvested first. For example, corn grown with a shorter plant would be harvested first, then peanut or sweet potato would be harvested later. Another pattern would be planting two tall crops with different growth rates. In relay inter-crops, different planting dates are used so that one crop might mature sooner.

Examples of Inter-crop Systems

Corn and Soyabean Mixed Intercrops

Canadian researchers (Martin et al., 1987) worked with several corn-soyabean inter-crop-seeding rates to determine their economic advantages as silage. Pure stands of corn and soyabeans were grown for comparison planted at 24,000 corn seed per acre or soyabeans planted at 200,000 seed per acre. Results showed that inter-crops were more cost effective than pure stands over both years the study was conducted. The study featured five experimental inter-crop seeding rates with two planting arrangements (alternate and within the row).

Strip Cropping corn/soyabeans/small grains

Iowa farmer Tom Frantzen strips as a crop rotation in one field. His rows are oriented generally east and west on the contour. His 1989 strip crop corn yields were 166 bushels per acre compared to 130 for his farm average. Stripped soyabean yields were 2 bushels lower than farm average. His oat yields were 109 bushels stripped and 100-bushel farm average. Tom was not surprised at the increase in corn yields. The outer strip rows captured more sunlight. His average corn border row yielded 198 bushels per acre next to the soyabeans and 177 bushels next to oats.

Limitations

- As with any crop, seed germination and seedling establishment is the most critical growth phase of the entire season. A good seedbed is needed to get a good stand. Delayed planting may reduce yield since crop development many not coincide with the optimal seasonal growth periods.
- Planning fertilization for intercrops can be challenging, as the full needs of both crops must be met. Generally, there is little information available on how to go about this. One possibility would be to ask for soil test results for each crop separately, then formulate a recommendation that will cover the needs of both crops to be grown. Such recommendations are generally 10% to 30% higher than rates for individual crops.
- As with any crop, also accounting for residual or carryover fertility from past crops saves money. Carryover fertility from inter-crops may well be lower than that of pure stands because of the two crops having different root types and feeding habits.
- Weed and pest control needs in inter-crops will likely be different than in pure stands. Some disease incidence, such as soyabean or mung bean rusts may increase, when aggravated with high corn populations and overfertilization. Any disease or pest that

prospers under shady conditions could increase, when grown under a taller crop, such as corn or sunflowers. In many cases, insect pests are lower when two or more crops are grown together.

Advantage and inference

Inter-cropping as a technique reduces risks in crop production; if one member of an inter-crop fails, the other survives and compensates in yield to some extent, allowing the farmer an acceptable harvest. Pest levels are often lowered in inter-crops, as the diversity of plants hampers movement of certain pest insects and in some cases encourages beneficial insect populations.

References

1. Anon. 1990. Strip inter-cropping offers low-input way to boost yields. Sensible Agriculture. May. p. 7-8.
2. Anon.2011-12. Annual Report, 2011-12. Indian Grass and Fodder Research Institute, Jhansi.
3. Anon. 2012. Annual Report-Kharif, 2012. AICRP on Forage Crops, IGFRI, Jhansi.pp.95-125
4. Anon. 2013. Annual Report-Kharif, 2013. AICRP on Forage Crops, IGFRI, Jhansi.pp.64-65.
5. Anon. 2017. ZREAC Report-Kharif, 2017. Agricultural Research Station, SKRAU, Bikaner.
6. Francis, R. and D.R. Decoteau, 1993. Developing an effective southernpea and sweet corn intercrop system. Hort Technology. Volume 3, Number 2. p. 178-184.
7. Grossman, Joel and William Quarles. 1993. Strip inter-cropping for biological control. IPM Practitioner. April. p. 1-11.

❑❑❑

Chapter 14

Organic Farming Scope, Prospects and Practices in Seed Spices

Dr. G. Lal
Director, ICAR-National Research Centre on Seed Spices, Ajmer (Rajasthan)

Introduction

The production of seed spices is concentrated mainly in the arid to semi-arid regions covering Rajasthan, Gujarat and Madhya Pradesh. A small group of farmers in these States have started organic cultivation but official information regarding its production, area and export is available a meager. The organic production of seed spices crops have not gained much momentum due to lack of scientific information on suitable production technology, appropriate plant protection measures and specific varieties for organic production. The growth of organic production of seed spices is linked with the the processing units.

Export Scenario

India has around 2.5 times rise in export since 2000 both in volume and value. It is on World No.1 in exporter of Cumin, Coriander, Fenugreek and Fennel Approx. 2,16,870 tonnes of seed spices export of value 2618.98 Crore INR (2015-16).

Organic Agriculture

The organic farming in real sense envisages comprehensive management approach to improve the health of underlying productivity of the soil (Palaniappan and Annadurai, 1999). Earlier Lambkin et al. (1999) mentioned that organic agriculture is a production system which avoids or largely excludes the use of synthetic compounded fertilizers, pesticides, growth regulators and livestock feed additives.

The indiscriminate and excessive use of chemicals during green revolution period has put forth a question mark on sustainability of agriculture in the long run. Organic farming appears to be one of the options for sustainability. The year 2000 is very important year for India from organic point of view.

Four major happenings were made during the year 2000

1. The Planning Commission constituted a steering group on agriculture who identified organic farming as National challenge and suggested it should be taken in the form of a project as major thrust are for 10th plan.
2. The group recommended organic farming in NE Region, rain-fed areas and in the areas where the consumption of agro chemicals is low.
3. The National Agricultural Policy (2000) recommended promotion of traditional knowledge of agriculture relating to organic farming and its scientific up gradation.
4. The Department of Agriculture and Co-operation (DAC), Ministry of Agriculture constituted (2000) a Taskforce on organic farming under the chairmanship of Shri Kunwar Ji Bhai Yadav and this taskforce recommended promotion of organic farming.
5. The Ministry of Commerce launched the National Organic Programme in April 2000 and (APEDA) is implementing the National Programme for Organic Production (NPOP) Gouri (2004).
6. Under the NPOP, documents like National standards, accreditation criteria for accrediting inspection and certification agencies.
7. Accreditation procedure, inspection and certification procedures have been prepared and approved by National Steering Committee.

Organic seed spice production

World over, people are becoming more conscious of the health problems due to consumption of agricultural produce with pesticide residues. Adverse impact of modern agriculture on the environment. Organic spices are highly remunerative because of higher demand in domestic market in metro cities and for export earnings. The quality of organic seed spices is comparatively higher. Organic spices are safe, nutritive and better for human health. Organic production technology is eco-friendly and compatible with farming situation. Fertility status of the soil is maintained over a longer period of time. It helps in enhancing availability of nutrients to plants. Organic farming practices are near to nature hence, dependency on external inputs is reduced. It is believed that organically produced seed spices may be more resistant to pest and diseases as compared to chemical farming.

Main principles of organic farming

To avoid all pollution forms that may result from agricultural techniques. To work as much as possible within a closed system of resource utilization. To reduce the use of fossil energy. Organize enterprise production for efficient use of resources. Cropping system, Crop rotation, inter-cropping, mixed cropping & relay cropping. Maintain soil fertility by use of renewable resources. Diversification of cropping systems for optimum production. Aims for optimum nutritional value of staple food. Strive for equitable relationship between those who work & live on the land. Decentralized structure for processing, distribution and marketing.

Impact of organic farming on yield of seed spices

- Diversion of existing farming system to organic way of seed spice production has certain impact on level of production.
- The impact of conversion to organic farming differs situation to situation and farmer to farmer.
- In intensive farming system, organic farming decrease yield, the range of decrease in yield depends on intensity of external input used before conversion and type of farming situation.
- In irrigated lands and well-endowed water regions conversion to organic farming leads to almost identical yields.
- In traditional rainfed agriculture with very low external inputs organic agriculture has shown potentials to increase in yield.
- Studies on other crops showed that under abiotic stress it gives higher yield over conventional farming.
- Organic farming have less long-term yield variability.

General Guidelines for organic seed spice production:

- Minimum area 0.40 ha is the eligibility requirement for organic spices growers.
- Applicant should possess a valid organic certificate.
- The entire farm should be converted to organic to avoid contamination.
- Crop rotation involving a legume in the cycle should be followed.
- A period of three years is required as conversion period for organic cultivation.
- In case of cultivation on virgin lands, the conversion period can be relaxed.
- Well rotten farmyard manure or compost @ 10-12 t/ha for fennel and 4-5 tonnes may be used for coriander, cumin and fenugreek as basal application.
- Aphids are major problem in case of seed spices crops, regular surveillance is necessary for effective management of this pest.
- Restricted use of Bordeaux mixture 1.0% for control of root rot in fenugreek, wilts in cumin and coriander, blights in fennel and cumin can be practices.
- Dusting of sulphur as a restricted use can be done for control of powdery mildew.

Components of organic farming

The major components of organic farming employed for management of nutrient, weed insect pest and diseases in organic seed spices production are:

- **Organic manures** like FYM, sheep manure, crop residues, poultry manure, oilcakes and other farm wastes, composts, coir pith compost, etc.
- **Green manuring** by Crotoleria juncea, Sesbenia spp and other legumes should be used for green manuring. Dhaincha is better in saline soils.
- **Vermi-compost** is also becoming popular now-a-days.
- **Bio-fertilizers :** Natural fertilizers containing carrier based micro-organism viz., Rhizobium, Azotobacter, Azospirillum, Blue-green

Alae, Azolla, Mycorrhizae and Phosphobacteria (Pseudomonas striota and Bacillus megathirum and VAM.

- **Biological agents** for controlling insects.
- **Weed management :** By Ecological, crop husbandry, mechanical, soil solarisation biological method and use of bio herbicide like De-vine, collego, bipolaris and bidophos may be used in organic farming.

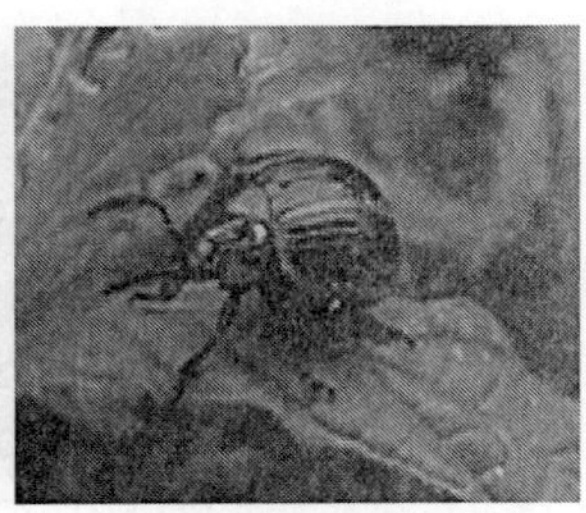

Technologies for organic seed spice production:

1. Coriander (Coriandrum sativum)

Coriander is an important major seed spice occupying first position in terms of production and second in respect to area and values among the major seed spices. Organic production of coriander is an very important aspect for enhancing its acceptance as organic product for better health and higher prices to farmers. The specific technologies are available for organic coriander production.

Technologies for organic coriander production

- Application of 10 to 15 tonnes vermi-compost in coriander significantly increased the concentration of trace elements (Fe, Mn) in leaves and Zn content also increased even with 5 tonnes vermin-compost (Reddy et al., 1993).
- The Good herbage yield of 6067 kg ha^{-1} obtained in coriander with 15 t ha^{-1} of vermi compost, whereas the highest seed yield of 1314 kg ha^{-1} was obtained with 20 t ha^{-1} of vermi compost (Vadiraj et al., 1998).
- The use of bio-inoculants Azospirillum or Azotobacter as seed treatment and soil treatment in combination with 5 t sheep manure ha^{-1} give higher seed yield under semi arid conditions (Vashishtha and Aishwath, 2008).
- The eco-friendly technique of using 5% onion leaf extract as foliar spray three times can also protect the plants from powdery mildew as revealed from studies at Coimbatore.
- The fusarium wilt incidence in coriander can be better managed by seed pelleting with Trichoderma viride (4g/kg of seed) plus neem cake application (150kg/ha) reduces wilt incidence (AICRPS recommendation).

2. Cumin (Cuminum cyminum)

Organic production of cumin is an additional feather in the cap to enhance its rate and organic cumin is better for consumption without any negative effect on health of human beings as well as to environment. The technologies available for organic cumin production are:

Cumin

Fennel

- Use of Azospirillum or Azotobacter in combination with 5 t/ha sheep manure/ha is suitable for organic cumin production.
- Use soil amendments like castor or mustard cake, poultry manure @ 2.5 t/ha before sowing for control of wilt.
- The crop rotation like cluster bean- cumin, cluster bean-wheat and cluster bean-mustard are better for management of cumin wilt.
- Use of Fusarium wilt resistant variety GC-3 and GC-4.
- Late sowing minimizes incidence of blight.
- The technique of soil solarization may be used for control of weeds, disease and pests.

3. Fennel (Foeniculum vulgare Mill)

Fennel is an important major seed spices crop playing very important role in increasing income of farmers growing vegetables because it has been instrumental seed spices for intercropping. Organic production of fennel is an important issue to fennel growing farmers for maintenance of health of human beings and environment along with sustainability in production.

Technologies for organic fennel production

- Use of bio-inoculant Azospirillum or Azotobacter as seed treatment and soil application with 10 tonnes/ha sheep manure is better for higher yield (AICRPS recommendation).
- Ramularia blight should be controlled with incorporation of Trichoderma viridae @4g/kg of seed.

4. Fenugreek (Trigonella foenium graecum)

Fenugreek is also an important major seed spices crop playing very important role as spice and medicine. Organic production of fenugreek is important for maintenance of soil fertility and health of human beings along with sustainability in production.

Technologies for organic fenugreek production

- Application of 7.5 t/ ha sheep manure with inoculation of seed by Rhizobium meliloti is recommended for good plant growth seed yield.
- The combined application of Azotobacter with Rhizobium and Pseudomonia striata improved the growth, yield and nodulation in fenugreek (Parakhia et al., 2000).
- To control damping off, prophylactic application of 1% Bordeaux mixture.
- To control aphids in fenugreek, application of fish oil, rosin soap or neem seed kernel extract (3%), neem oil or tobacco decoction (0.05%) is effective (AICRPS recommendation).

Organic Fenugreek

Crop protection by organic means

- The plant protection measures under organic farming system should involve emphasis on crop management practices time of sowing, balanced nutrition, crop rotation, green manuring, etc. for reducing the incidence of disease and pests, selection of resistant varieties and adoption of bio-control measures.
- In a study carried out at NRCSS, seven different phytological extracts/biological agents viz., Neem oil (2%), Karanj oil (2%), combination of Neem oil and Karanj oil, garlic extract (1%), combination of sulphur compound and isothiocynate (1%), Ajwain extract (5%), and nigella extract (5%) were used after one month of germination of the crop. It is revealed that foliar spray of bio-fungicides Neem Oil + Karanj oil exhibited maximum yield (2495 kg/ha) in coriander with effective control on the diseases.

Management of soil borne diseases in seed spices

Organic amendments: Application of mustard residue (2.5 t/ha) + mustard cake (0.5 t/ha) + neem cake (0.5 t/ha) as soil amendment with one irrigation during hot summer.

Bioagents : Application of Trichoderma viride or Aspergillus versicolar as seed treatment @ 10 g/kg and soil application @ 2.5 kg/ha at the time of sowing.

Soil solarization : Soil solarization for 20-30 days in hot summer, followed by seed treatment (10 g/kg seed) and soil application (2.5 kg/ha) of T. viride at the time of sowing.

Bio-agents and botanicals developed

Developed 7 bio-control agents viz., entomopathogens: Metarhizium anisopliae, Verticilium lacani, Beaveria bassiana, Soil antagonist: Trichoderma viridae, Trichoderma harzanium, Pseudomonas fluorescens, nematopathogens : Paecilomyces lilacinus and botanicals : Allyl iso thi cynate, Sulphur Extract of Karanj, Tumba fruit extract and Kareel extract.

Constraints in adoption of organic farming in seed spices:

- Lack of awareness and marketing problems of organic produce.
- High input costs and inadequate supporting infrastructure.
- Shortage of bio-mass as nutrient source and marketing problems of organic inputs.
- Lack of financial support and low yields.
- Inability to meet the export demand.
- Vested interests and lack of quality standards for bio-manures.

Promotion of organic seed spices production

- Dissemination of organic farming in holistic manner.
- Integrated efforts of supporting agencies.

- Encouragement of decentralized input supply.
- Adoption of improved methods of composting.
- Awareness and capacity building.
- Subsidy on organic inputs and promotion of high value crops.
- Development of organic clusters/villages.

Future strategies to promote organic seed spice production

- Development of varieties suitable for organic farming which are resistant to biotic and abiotic stresses.
- Work out economically viable cultural practices, viz., cropping pattern, crop rotation, inter-cropping, solarization, crop mixing, mulching for organic farming.
- Development of organic production technology using bio-fertilizers and bio-control agents and soil amendments to control soil pathogens.
- Formulation of forecasting models for blight, powdery mildew, aphids in organic seed spices crops and use of effective and eco-friendly control measures for the disease and pests including storage pests.

Conclusion

- The organic farming is an integrated approach starting from crop raising to harvesting, post harvest handling, processing and storage.
- Separate set of implements, equipment and machinery is needed for organic seed spice production to maintain prescribed standard of quality and to avoid contamination with conventional production.
- The ploughing, threshing and grading equipment should be separate for organic system of production to ensure proper quality.
- Organically produced seed spice improves the health of the consumers and wealth to the producer, thus is the boon to industry. Best method for quality and eco-friendly sustainable production of seed spices.

Chapter 15

Transition to Organic Farming in Rainfed Agriculture : Status, Strategies and Issues

Dr. S.K. Sharma
Director, Centre for Advanced Faculty Training on Organic Farming

Evolution of Organic Agriculture

In 1911, farmers of Forty Centuries by F.H. King. Observations in Asia. 1927 : The Living Soil by Lady Eve Balfour. What else was going on in the soil. In 1940 : An Agricultural Testament by Sir Albert Howard. Work in India with composting and the effect of compost on soils and food quality. In 1945 Pay Dirt, J.I Rodale American, the soil is alive. In 1900's: Understanding of soil chemistry and processes influencing nutrient availability. The value of legumes was well recognized (Hopkins, 1911).

1960's and 1970's Alternative view to conventional agriculture lead sustainable agriculture. The concepts of sustainable agriculture are based on developing an agriculture system that will protect soil quality and health for the long-term.

Methods or Schools of Organic Farming

1. Biodynamic Agriculture
2. The Rodale's Alternative Agriculture
3. Bio-intensive Farming
4. Natural Farming
5. No Work Method
6. Veganics

Biodynamic

Rudolph Steiner in Germany during the 1920's. His basic theory was that nature is a mystical, spiritual thing, and to garden it well, one must treat it as a whole entity and system, both physical and spiritual. Farming practices are related to lunar and planetary phases.

The Rodales Alternative Agriculture

J.I. Rodale in early 1940's. Grew own food by own resources. Chemicals used as potential poisons did not make sense to him. Composting methods of Sir Albert Howard in India (An Agricultural Testament), the work of Lady Eve Balfour (The Living Soil), and F. King (Farmers of Forty Centuries). Emphasis on human health and organic gardening Son, Robert Rodale. The family developed the Rodale Institute, Organic Farming and Gardening Magazine, and Rodale Press.

Bio-intensive Farming

How to Grow More Vegetables Than You Ever Thought Possible on Less Land Than You Can Imagine (1974, 5th edition in 1995). John Jeavons. Originally by Alan Chadwick, an English horticulturist who combined French Intensive gardening methods, such as double digging with Biodynamic methods. On small plots of land around the world. The organization "Ecology Action" works to promote bio-intensive methods and how they can be used to develop food supply for local communities.

Natural Farming

Masanobu Fukuoka in Japan (One Straw Revolution, 1978). Building soil with regular surface applications of organic matter decompose on the surface as opposed to composting in piles. Little or no cultivation. Weeds are controlled by applications of more mulch or organic matter.

No Work Method

This method is very similar to the Natural Farming method. 1971 "The Ruth Stout No-Work Garden Book" (Rodale Press, Emmaus, Penna, 1971). The emphasis is on lots of organic matter for mulch to control weeds and provide nutrients. Worms and other organisms are allowed to do the work of getting nutrients to the roots.

Veganics

Term 'veganics' as opposed to organics. Nutrient management on only plant derived nutrients and organic matter as opposed to making use of manures and other animal products. Not widely used.

Phases of Organic Agri. Movement

Factors

The ill-effects' and the mindset of consumers, High premium for health, Policy-makers restoration of soil health and Generation of rural economy apart from creating better environment are the factors for organic farming.

IFOAM definition

IFOAM (1996) Basic Standards : The "Principle Aims of Organic Agriculture and Processing" principles and ideas:

- To produce food of high nutritional quality in sufficient quantity;
- To interact in a constructive and life enhancing way with all natural systems and cycles;
- To encourage and enhance biological cycles within the farming system, involving micro organisms, soil flora and fauna, plants and animals;
- To maintain and increase **long-term fertility** of soils;
- To promote the healthy use and proper care of **water**, water resources and all life therein;
- To help in the **conservation** of soil and water;
- To use, as far as is possible, **renewable resources** in locally organized agricultural systems.

IFOAM Principles

- To work, as far as possible, within a closed system with regard to organic matter and nutrient elements;
- To work, as far as possible, with materials and substances which can be reused or recycled, either on the farm or elsewhere;
- To give all livestock conditions of life which allow them to perform the basic aspects of their innate behaviour;

- To minimize all forms of pollution that may result from agricultural practices;
- To maintain the genetic diversity of the agricultural system and its surroundings, including the protection of plant and wildlife habitats;
- To allow everyone involved in organic production and processing a quality of life conforming to the UN Human Rights Charter, to cover their basic needs and obtain an adequate return and satisfaction from their work, including a safe working environment;
- To consider the wider social and ecological impact of the farming system;
- To produce non-food products from renewable resources, which are fully biodegradable;
- To encourage organic agriculture associations to function along democratic lines and the principle of division of powers; entire organic production chain;
- To progress towards an, which is both socially just and ecologically responsible.

Concept of Organic Agriculture

A labelling term for a product. Produced in accordance with certain standards during food production, handling, processing and marketing stages, and certified by a duly constituted certification body or authority. Process claim rather than a product claim. Not necessarily mean that the foods produced are healthier, safer or all natural. Follow the defined standard of production and handling. Not exempt from compliance with general regularity requirements, such as food safety regulations, pesticide registration, general food and nutrition labelling rules, etc. Distinct practices for organic farm and insect pest & disease and food production systems.

Principles of Organic Agriculture

The fundamental rules-Use as a whole. Ethical principles to inspire action. Roots from which organic agriculture grows and develops. Humankind's most basic activities (History, culture and community values) are embedded. The way people treat soils, water, plants and animals in order to produce, prepare and distribute food and other goods.

Period of Struggle	• 1924 – 1970 • Economic Struggle • Basic work on Principles & Causes
Period of Labels	• 1970 – 1990 • Consumer awareness • Organic product demand • Discussion on need of labels & standards
Period of Acceptance	• 1990 onwards • Acceptance of standards & certification • Market increase • Legal declaration

Four Basic Principles

1. Principle of Health

Health of individuals and the health of ecosystems. Wholeness and integrity of living systems. Absence of illness, but the maintenance of physical, mental, social and ecological well-being. Immunity, resilience and regeneration. Health of ecosystems and organisms from the smallest in the soil to human beings. Products that contribute preventive health care and well-being.

2. Principle of Ecology

Ecological processes and recycling. Nourishment and well-being: Ecology of the specific production environment. Crops: Living soil: Animals: Farm ecosystem; Fish and marine organisms: Aquatic environment. Production systems should fit the cycles and ecological balances in nature. These cycles are universal but their operation is site specific (Ecology, culture and scale).

3. Principle of Fairness

Equity, respect, justice and stewardship. Human relationships and fairness at all levels and to all parties. A good quality of life, and contribute to food sovereignty and reduction of poverty. Animals provided with the conditions and opportunities of life that accord with their physiology, natural behaviour and well-being. Natural and environmental resources use socially and ecologically just and trust for future generations. Production, distribution and trade systems are open and equitable and account for real environmental and social costs.

4. Principle of Care

Living and dynamic system. Enhance efficiency and increase productivity upto well-being level. New technologies with care & complete understanding of ecosystems. Precaution and responsibility. Science is necessary but scientific knowledge alone is not sufficient. Practical experience: Valid solutions (Tested by time). Prevent significant risks in technologies and rejecting unpredictable ones (GE).

Economic viability of agriculture

Agriculture, once a mode of life, has become a made of production (Krimsky & Wrubel, 1996). Income generation was only a fringe benefit. Changes in the mode of agricultural production reflect concurrent changes in other aspects of social & cultural life. New interactions in production system, so everything has to be assigned a market value.

Measures of sustainable farm

Yield, Long-term effect and Minimum risk are the three measures of sustainable farm.

Literature review – NARS system

Those dealing with the problem of agriculture and suggesting alternative systems, such as those expounding the principles and practices of those which study the different elements, such as bio-inputs on comparative yields.

Acceptance of Organic Agriculture in India

Second generation problems of GR, Rice-Wheat Multi-nutrient deficiency on animal production, LTFE results, 10 mt gap in NSD, Increasing cost of National & International trade effect, Agri-Export opportunities, Alternative Energy Source and Pressure on Resources.

Year 2000 : Watershed Year in Indian Organic History

Steering Group on Agriculture set (By planning commission in 2000). Organic farming as a national challenge & a major thrust area for 10th plan- (NE, Rainfed, low agro-chemical consumption). The National agricultural Policy (2000)-Recommended for promotion of traditional knowledge of agriculture relating to of & its scientific upgradation. DAC (MOA)-Task force on Organic Farming (2000)

Rainfed Agriculture : Low Input Farming

Low external input or no external input farming integrated with livestock, particularly small ruminants. The average use of chemical fertilizers in rainfed areas based on a survey of non-irrigated SAT districts was found to be 18.5 kg as against 58 kg in the irrigated districts (Katyal and Reddy, 1997). Upto 30% of the rainfed farmers in many remote areas of the country do not use chemical fertilizers and pesticides. Resource poor farmers are practicing organic farming by default. The Government of India taskforce on organic farming and several other reviewers have identified rainfed areas and regions in north east as more suitable for organic farming in view of the low input use.

Sustainability of Yield of Dryland Crops

The All India Co-ordinated Research Project on Dryland Agriculture and related projects during the last 25 years, it is evident that chemical fertilizers have significantly contributed to improved productivity in rainfed crops, even in areas where legumes are part of the cropping systems. However, long-term data suggest that even in drylands, sustainability of higher yields over a period of time is possible only when optimum nutrients are supplied through organics or a combination of organics and chemical fertilizers but not when supplied as chemical fertilizers alone (Hegde, 1988).

Essential Components of Organic Rainfed Organic Farming

Adoption of soil and water conservation measures, Mulching or mulch cum manuring. Residue management, Green leaf manuring. Cover cropping, Vermicomposting in particular has been accepted by the rainfed farmers and gained momentum during the last 5 years, Beejamrut, Jeevamrut and Whapsana.

Harmony b/w Organic Farming & Rainfed Farming

The use of FYM or other organic nutrient sources during aberrant rainfall years in particular have an additional advantage of protecting the crop from drought besides the nutritional benefits, so critical in drylands. While there is no contradiction between these established rainfed farming technologies and the objectives of the organic farming, the only issue will be the labour and capital intensive nature of some of these technologies and its ultimate impact on the cost of production.

The Key Issues

- Clearly defining our goals. Mind set change on the concept of inputs and crop management.
- Building on traditional knowledge and INM/ IPM technologies as starting points
- Look for sustainable yields and environmental quality at similar or lower production costs
- Integrate soil and water conservation practices-the bedrock of organic farming
- Best opportunity lies in cropping systems (to take care of nutrient and pest management).
- Labour availability is a myth. Mechanization of production and application of on farm generated inputs (biomass)/composting is a dire need.

Conclusion

- Organic Farming is a dynamically evolving system as is conventional farming.
- Comparison of two systems may change over time (New regulation, GMO, stringent environmental regulation, new technologies)
- Organic Farming can currently be seen as a role model for creating an environmental friendly farming system.
- Organic Farming can have indirect influence on conventional farming whose effects might even go beyond the direct effects of Organic Farming on the environment.
- To keep this role model alive & developing in arid semi-arid areas it further, research & other measures need to be strengthened.

❑❑❑

Chapter 16

Precision Agriculture: Key to Next Agriculture Revolution

V.M. Mayande

Former Vice Chancellor, Dr. PDKV, Akola and Member VII QRT-AICRPDA

Precision agriculture (PA) is a method of farm management that allows the farmer to produce more efficiently, thereby realizing gains through economical use of resources. An important effect of precision farming is the high environmental benefit from using chemical treatments only where and when they are necessary. This promotion of environmental stewardship is a key component of the new attitudes in agriculture. This is an approach to farm management that uses information technology (IT) to ensure that the crops and soil receive exactly what they need for optimum health and productivity. The goal of PA is to ensure profitability, sustainability and protection of the environment. PA is also known as satellite agriculture and as needed site-specific crop management (SSCM).

Historical Perspective

Precision agriculture is a key component of the next wave of modern agricultural revolutions. The first agricultural revolution in western countries came through increased mechanization in 1900-1930. Each farmer produced enough food to feed about 26 people during this time. This prompted the Green Revolution with genetic modification, which led to each farmer feeding about 155 people. It is expected that by 2050, the global population will reach about 9.6 billion, and food production must effectively double from current levels in order to feed every mouth. With new technological advancements and revolution of precision farming, each farmer will be able to feed 265 people on the same acreage.

Precision Farming: Indian Perspective

India missed the agriculture revolution through mechanization as large population employed in agriculture will be jobless. This proved to be a myth as mechanized states in India created more employment in agriculture. New crop genotypes brought the first green revolution in India, however, its impact fading away due to excessive use of chemical inputs leading to environmental and resource degradation. It is now realized that the next agricultural revolution in India will come through Mechanization and Precision Farming. Mechanization is fast picking up in India and precision farming is just at beginning phase. India has a great strength in satellites, electronics & software technologies and now fast spread of communication devices. With these assets at disposal, precision agriculture in India is an achievable goal.

Precision Agriculture Phases

The precision agricultural revolution in first phase is through satellite and aerial imagery, weather prediction, variable rate fertilizer application and crop health indicators. The second phase aggregates the machine data for even more precise planting, topographical mapping, and soil data. Precision agriculture aims to optimize field-level management by matching farming practices more closely to crop needs, environmental protection and economics by boosting competitiveness through more efficient practices. Precision agriculture provides a wealth of information to build up a record to improve decision-making, foster greater traceability, enhance marketing of farm products. Prescriptive planting is a type of farming system that delivers data-driven planting advice that can determine variable planting rates to accommodate varying conditions across a single field, in order to maximize yield.

Precision Agriculture controls

Precision agriculture relies upon specialized equipment, software and IT services. The approach includes accessing real-time data about the conditions of the crops, soil and ambient air, along with other relevant information, such as local weather predictions, labour costs and equipment availability. Sensors in fields measure the moisture content and temperature of the soil and surrounding air. Satellites and robotic drones provide farmers with real-time images of individual plants. Information from those images can be processed and integrated with sensor and other data to yield guidance for immediate and future decisions, such as precisely what fields to water and when or where to plant a particular crop. The control units integrate sensor data and imaging input with other data, provides ability to identify fields that require treatment and determine the optimum amount of water, fertilizers and pesticides to apply. This helps to avoid wasting resources ensuring that the soil has just the right amount of inputs for optimum health, reducing costs and controlling the environmental impact.

Precision Agriculture Process

Precision agriculture enables site-specific treatment which refers to treating the smallest possible area as a single element. Site-specific management calls for treatment of only those few isolated weed spots rather than spraying whole field. The treatment of each site is specific to the needs of each site, as determined by soil test data and crop scouting reports. The implementation of the GPS has enabled to accurately map the fields and narrow down site size to only a few metres' area. Soil testing before planting and the data analysis determines spatial variations in soil fertility variations. Planting with variable seeding rates across the field and use of variable rate application of fertilizers is determined by soil test data. Crop scouting is done to search for problems, such as weeds, pests, or diseases. During the harvesting process, variable yield results across the field can help the farmer discover if his management methods have been successful.

Yield Monitors

Yield monitoring is probably the most important cornerstone of precision agriculture. A method of instantaneous yield monitoring utilizes sensors in the combine to continuously log grain flow rate during harvesting, and the combine's speed. This data, combined with the GPS location for each data point, allows the creation of a yield map within a GIS. The yield map can compare to maps of soil test data, chemical application maps, and other information, resulting in a recommendation for the next year's site specific management plan. When layered in a GIS with associated maps, a yield map displays evidence of relationships between crop yield and field condition variables.

Precision Agriculture stages

Precision agriculture is usually done as a process to observe spatial variability in four stages:

1. Data Collection

Geo-locating a field enables to overlay information gathered from analysis of soils and residual nitrogen, and information on previous crops and soil resistivity. Geo-location is done in two ways: The field is delineated using an in-vehicle GPS receiver as the farmer drives a tractor around the field. The field is delineated on a base map derived from aerial or satellite imagery. The base images must have the right level of resolution and geometric quality to ensure that geo-location is sufficiently accurate.

2. Variables

Intra and inter-field variability may result from a number of factors. These include climatic conditions, soils, cropping practices, weeds and disease.

3. Strategies

Soil maps can be used to predict input requirement based on analysis of static indicators (soil, resistivity, field history) during the crop cycle. Information from static indicators is regularly updated during the crop cycle. Sampling, remote sensing approach can be used to plan strategy.

4. Implementing practices

Geo-locating a field enables to overlay information gathered from analysis of soils and residual nitrogen, and information on previous crops and soil resistivity. Geo-location is done in two ways: The field is delineated using an in-vehicle GPS receiver as the farmer drives a tractor around the field. The field is delineated on a base map derived from aerial or satellite imagery. The base images must have the right level of resolution and geometric quality to ensure that geo-location is sufficiently accurate. Intra and inter-field variability may result from a number of factors.

Predictive approach : based on analysis of static indicators (soil, resistivity, field history, etc.) during the crop cycle.

Control approach : Information from static indicators is regularly updated during the crop cycle by sampling, weighing biomass, measuring leaf chlorophyll content, weighing fruit, etc.

Remote sensing : Measuring parameters like temperature, humidity, wind or stem diameter is possible using Wireless Sensor Network.

Satellite remote sensing : Multispectral imagery is acquired and processed to derive maps of crop biophysical parameters, including indicators of disease. Airborne instruments are able to measure the amount of plant cover and to distinguish between crops and weeds. Decisions may be based on decision-support models (crop simulation models and recommendation models), but in the final analysis it is upto the users to decide in terms of business value and impacts on the environment. Application of crop management decisions calls for agricultural equipment that supports variable-rate technology (VRT), for example varying seed density along with variable-rate application (VRA) of nitrogen and phyto-sanitary products. Precision agriculture uses technology on agricultural equipment (e.g. tractors, sprayers, harvesters, etc.): positioning system (e.g. GPS receivers that use satellite signals to precisely determine a position on the globe); geographic information systems (GIS), i.e., software that makes sense of all the available data; variable-rate farming equipment (planter, harvester, etc.).

Precision Farming around the world

The concept of precision agriculture first emerged in the United States in the early 1980s. It was also at this time that the practice of grid sampling appeared. Towards the end of the 1980s, this technique was used to derive the first input recommendation maps for fertilizers and pH corrections. The use of yield sensors developed from new technologies, combined with the advent of GPS receivers, has been gaining ground ever since. Today, such systems cover several million hectares. One-third of the global population still relies on agriculture for a living. Although more advanced precision farming technologies require large upfront investments, developing countries are benefitting from mobile technology.

Economic and environmental impacts

Precision agriculture implies application of precise and correct amount of inputs like water, fertilizer, pesticides, etc. at the correct time to the crop for increasing its productivity and maximizing its yields. Precision agriculture management practices can significantly reduce the amount of nutrient and other crop inputs used while boosting yields. Farmers, thus, obtain a return on their investment by saving on water, pesticide, and fertilizer costs. The second, larger-scale benefit of targeting inputs—in spatial, temporal and quantitative terms—concerns environmental impacts. Applying the right amount of inputs in the right place and at the right time benefits crops, soils and groundwater, and thus, the entire crop cycle. Consequently, precision agriculture has become a cornerstone of sustainable agriculture, since it respects crops, soils and farmers. Sustainable agriculture seeks to assure a continued supply of food within the ecological, economic and social limits required to sustain production in the long-term. Precision agriculture, therefore, seeks to use hi-tech systems in pursuit of this goal. Precision agriculture can help farmers in developing countries like India.

Emerging technologies

Robots

Self-steering tractors have existed for some time now, as John Deere Equipment works like a plane on autopilot. The tractor does most of the work, with the farmer stepping in for emergencies. Technology is advancing towards driverless machinery programmed by GPS to spread fertilizer or plow land. Other innovations include a solar powered machine that identifies weeds and precisely kills them with a dose of herbicide or lasers. Agricultural robots already exist, but advanced harvesting robots are being developed to identify ripe fruits, adjust to their shape and size, and carefully pluck them from branches.

Drones and satellite imagery

Advances in drone and satellite technology benefits precision farming because drones take high quality images, while satellites capture the bigger picture. Light aircraft pilots can combine aerial photography with data from satellite records to predict future yields based on the current level of field biomass. Aggregated images can create contour maps to track where water flows, determine variable-rate seeding, and create yield maps of areas that were more or less productive.

Precision Agriculture in Livestock farming

The network of physical objects outfitted with electronics enable data collection and aggregation. It comes into play with the development of sensors and farm-management software. Innovations are not just limited to plants-they can be used for the welfare of animals. Cattle can be outfitted with internal sensors to keep track of stomach acidity and digestive problems. External sensors track movement patterns to determine the cow's health and fitness, sense physical injuries, and identify the optimal times for breeding. All this data from sensors can be aggregated and analysed to detect trends and patterns.

Summary

Precision Agriculture Technology is increasingly practiced in developed countries using advanced tools and resources. It is recognized that next agriculture revolution has to come through precision farming. India has strength in terms of well-developed IT industry, Electronic industry, Satellite resources, agriculture tools and machinery. Integrated research and development program on precision agriculture in India is necessity to double the productivity from limited land and water resource. Precision Agriculture and Mechanization are keys to second agriculture revolution in India.

Bibliography

1. https://en.wikipedia.org/wiki/Precision_agriculture
2. https://www.alliedmarketresearch.com/
3. http://www.aef-online.org/
4. http://whatis.techtarget.com/definition/precision-agriculture-precision-farming
5. Betsy Pfister December 2, 1998:https://directionsmag.com/author/1013.

❑❑❑

Chapter 17

Farm Mechanization

Dr. O.P. Pareek

Senior Scientist and Head (Agricultural Engineering), KVK, Bhilwara

Introduction

Farm mechanization means the use of machines & technologies in the Agriculture sector for improving productivity. Mechanization may be either partial or complete. It is partial when only a part of the farm work is done by machine. When animal or human labour is completely dispensed with by power supplying machines, it is termed as complete. Mechanization of agriculture has two forms-mobile mechanization and the stationary types of mechanization. The former attempts to replace animal power on which agriculture has been based for very many centuries; while the latter aims at reducing the drudgery of certain operations which have to be performed cither by human labour or by a combined effort of human beings and animals.

State of Mechanization

Agriculture system has undergone changes in terms of cropping system, type of power sources used and application of inputs to achieve high level of productivity. Shortage of labour and high labour wages of the main factor which strongly propel mechanization. Engineering inputs in agriculture for effective utilization of inputs and thereby increasing the productivity and adding values to the produce. To meet the future food requirements, the farmer may require better production and processing technologies. Value addition and packaging technologies for promotion of agri-business and commercial agriculture.

Different ways of Mechanization

- By introducing the improved agriculture implements for small size holdings.

- By introducing the small tractors, tractor drawn machines, power tillers on medium size land holdings to supplement existing sources.
- By introducing the large size tractors and machineries for large farm holdings (> 10 ha).

Reasons for farm Mechanization

The present agriculture scenario is as under:

- Cultivable land 159.20 million ha.
- Farm holders 137.80 million.
- Average land holding size 1.16 ha.
- Medium to large farm holders 32%
- Small and marginal farm holders 68%

Small farmers due to their limited resources contribute to low productivity of land, as they depend on traditional equipments and methods of cultivation. Because of the low productivity of their lands, they also use low amount of crop inputs and do not adopt high yielding varieties of seed. Small and low farm holders have important role to play in higher agricultural production.

Scope of Mechanization due to:

1. Improved irrigation facilities in the area.
2. Introduction of high yielding varieties.
3. Introduction of high dose of fertilizer and pesticides for different crops.
4. Introduction of new crops in different parts of the country.
5. Multi-cropping system and intensive cultivation.

Indian Agriculture

India has net sown area of 140 million ha (42.6%), agricultural workers of around 263 million and provides livelihood to about 60% of the population which contributes 14% to the Gross Domestic Product (GDP). Mechanization can approach in Indian agriculture by improved input use efficiency and enhanced cropping intensity, reduced cost of production and drudgery of agriculture workers and by enhancing farm power supply and maintain a socially desirable mix of human labour, draught animal power and mechanical power.

Farm Mechanization in Rajasthan

Net Area Sown is the most important category and half of the total report area is under agriculture operation. Largest state covering 10.5% of geographical area. 5.5% population of the country. 2/3 population depends on agriculture. 10% animal population of the country. Only 1% of water resources of the country. 83% of available water used for irrigation. 35% of cropped area irrigated. 70% area irrigated through wells & tubewells. 27% irrigated through canals. 10.6 Lac ha saline and sodic soils.

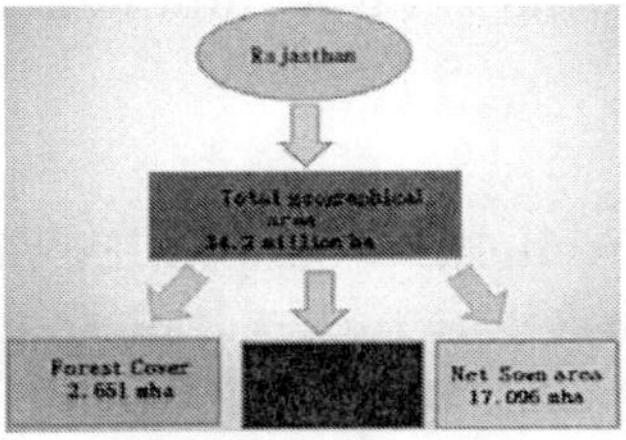

Farm mechanization saves Time, Labour, cuts down crop production costs, reduces post-harvest losses and boosts crop output and farm income. It creates strong correlation between farm mechanization and agricultural productivity. It is found that states with greater availability of farm power show higher productivity.

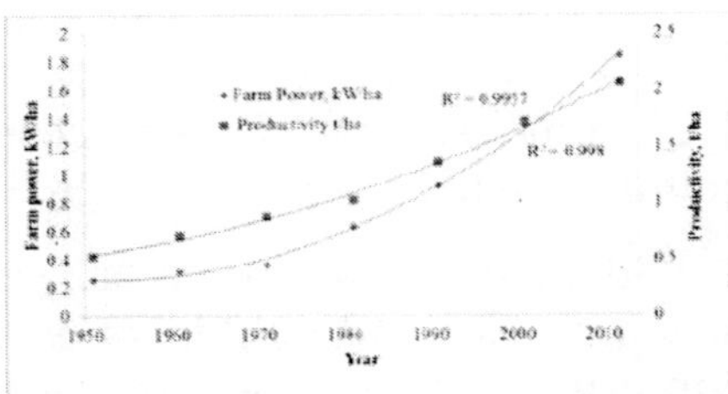

Fig. 1: Trend of farm power and agricultural productivity from 1951 to 2012

Farm Power Intensity Region

India has available farm power of about 1.80 KW/ha and available farm power in Rajasthan of around 0.7 KW/ha. It includes Human power (0.05 KW/ha), Animal power (0.09 KW/ha), Stationary power

(0.28 KW/ha) and Tractor power (0.58 KW/ha). Barmer, Bikaner, Churu, Jaisalmer, Jalore, Jhunjhunu, Jodhpur, Pali, Sikar, Tonk and Hanumangarh districts have power less than 1 KW/ha of net sown area . Alwar, Bharatpur, Dausa, Sri Ganganagar, Kota and Udaipur have more than 2KW/ha.

Opportunities

It creates a vast opportunities in Rajasthan as entrepreneurship development for custom hiring in Rajasthan under Farm machinery and Agro-processing equipment, developed agro-processing centres in Rajasthan to reduce post harvest loss reduction and value addition at the production. It also establishes a value chain for commercial supply. It will enhance agriculture by introducing micro-irrigation and adopting precision farming techniques.

Threats

It was found that sometimes its profitability is low in agricultural enterprises. Recent data concluded that migration of farmers from agriculture is increased. There is fragmentation and continuous reduction of operational holdings. It has analysed that with government intervention also, it has slow pace of R&D and commercialization. There is inadequate infrastructure back up for after sales support for farming equipments. Renewable energy technology still subsidy dependent. Unavailability of power and machinery at peak time.

References

1. Aderoba, A. 1987. A model for selective mechanization for the small farmer. Agricultural Systems, 25(3): 229–236.
2. Adrian, A. R., O. Akira, H. Nakashima, J. Miyasaka, and O. Katsuaki. 2007. Mechanization index and machinery energy ratio assessment by means of an artificial neural network: a Mexican case study. Agricultural Engineering International: the CIGR E Journal IX:1-21.
3. Baruah, D. C., and G. C. Bora. 2008. Energy demand forecast for mechanized agriculture in rural India. Energy Policy, 36(7): 2628–2636.

□□□

Chapter 18

Forages Management Under Drylands

S.M. Kumawat

Prof. Agronomy, Department of Agronomy, College of Agriculture

S K Rajasthan Agricultural University, Bikaner-334 006

Introduction

Drylands cover more than 60% of the earth's surface. Arid zones are described as a part of the drylands, and have the most severe edaphic-climatic conditions. The major distinguishing feature for defining and for planning for the arid zone is the low rainfall (below 500mm or Aridity Index <0.20) with more than 50% inter-annual variability. This makes a great difference in terms of the nature of the ecosystem, the socio-economic environment and the challenges for sustainability.

Of the total land area in the world, arid zones cover 18.8%. The hyper-arid zones account for 22.3% of the total arid zones (FAO, 1989). The hot arid regions of India lie between 24° and 29° N latitude, and 70° and 76° E longitude, covering an area of 31.70 million hectares, and involving seven states: Rajasthan, Gujarat, Punjab, Haryana, Andhra Pradesh, Karnataka, and Maharashtra. In total, 11.8% of the country is under a hot arid environment. The arid regions of Rajasthan, Gujarat, Punjab, and Haryana together constitute the Great Indian Desert, better known as the Thar Desert, which accounts for 89.6% of the total hot arid regions of India, and constitutes the principal hot arid zone of the country.

Landuse options that increase livelihood security and reduce vulnerability to climate change are necessary. Agroforestry is one type of landuse which improves the degraded natural base besides providing economic products to the community. Agroforestry is widespread in all ecological and geographical regions of India. Agroforestry systems vary enormously in their structural complexity and species diversity, their productive and protective attributes and their socio-economic dimensions. It is aimed to optimize the use of resources through recycling, internalize the input production, reduce the risk and conservation of natural resources.

Agro-forestry Systems

Farmers in the hot Indian arid zone grow arable crops in association with tree species(mainly Khejri). In fact, such integration of arable crops with trees in the farming systems is a unique, combined, protective-productive system that works on the principles of ecology, productivity, economics and sustainability. These systems are now generally referred to as agroforestry. Because most trees are drought resistant, they are still able to provide fuel, fodder, fruit and other products, when the crops fail, as frequent droughts are a common phenomenon.

According to India State of Forest Report (2013), Rajasthan (8,373sqkm)is the third state having maximum tree green cover are in agroforestry system next to Maharashtra (11,806sqkm) followed by Gujarat (11,591sqkm). Most dominant agroforestry system in western Rajasthan is Prosopis cineraria based, which cover about 47% of the total area (Table 1).

Table 1: Tree component involved in the various agroforestry systems in arid region of Rajasthan.

System	Surface Area (square km)	Per centage	Districts
Prosopis cineraria (P. spicigera)	158,647	46.3	Barmer, Bikaner, Churu, Sri Ganganagar, Sikar Jaisalmer, Jalore,Jodhpur, Junjhunu, Nagaur,
Ziziphusn-ummularia	94,440	27.6	Barmer, Bikaner, Jaisalmer, Jodhpur
Acacia niloticasubspindica	49,594	14.5	Sri Ganganagar, Jodhpur, Nagaur,Pali
Tecomellaundulata	22, 937	6.7	Barmer
Acacia tortilis	16,672	4.9	Sri Ganganagar
Total	**342,290**	**100.0**	

Fodder production potential of Agroforestry Systems

There are various agroforestry systems which provides fodder, fuel, food and fruits, such as agri-silviculture (crop + tree/animals), silvi-pasture (tree + pasture/+ animals), agri-horti (crop + fruit trees), horti-pasture (fruit trees + pasture/+animal), and agri-horti-silvipasture (crop + fruit trees + MPTS + pasture). Many Multipurpose tree species (MPTS) growing in agroforestry systems are useful as leaf fodder used for animal feed besides wood. Grazing animals with MPTS trees provide not only nutritious fodder but also shelter to the animals during bright and hot sunny day.

Major traditional agroforestry systems

On the basis of rainfall, four types of major traditional agroforestry systems have been identified (Table 2). Over 400 mm rainfall zones are pre-dominated with Prosopis cineraria-Acacia nilotica based system; between 300 and 400 mm rainfall zone P. cineraria based; between 200-300 mm rainfall zone Zizyphus spp. P. cineraria based; and in less than 200 mm rainfall zone Zizyphus spp. P. cineraria–Salvadora spp. based systems are followed. The tree/shrub density of the system forming species is also clearly governed by rainfall regime(Pandey, 2007). Data indicated that with decrease in rainfall, the density of woody component of the system decreased substantially and as well as the productivity of arable crops & grasses were declined.

Table 2: Traditional agroforestry systems in arid region of western Rajasthan

Rainfall zone (mm)	Agroforestry system	Trees/shrubs (Nos./ha)	Density of prominent species (%)
>400	P. cineraria - A. nilotica based	31.4	80.5
300-400	P. cineraria based	14.2	80.0
200-300	Zizyphus spp. - P. cineraria based	91.7	100.0 (91.7% Zizyphus spp.)
<200	Zizyphus spp. - P. cineraria – Salvadora spp. based	17.2	87.2 (65% Zizyphus spp.)

It was very interesting that yield of pearl millet, main cereal crop of the region, below the canopy of woody components in any system was not affected at all, rather in Zizyphus spp. - P. cineraria and Zizyphus spp. - P. cineraria - Salvadora spp. systems it was substantially increased below the tree canopies in comparison to open spaces. However, legumes, such as moth bean, cluster bean and mung bean exhibited slight yield reduction below the tree canopies, 5.0% (Zizyphus spp. - P. cineraria system) to 19.5% (P. cineraria system); 12.5% (Zizyphus spp. - P. cineraria - Salvadora spp. system) to 18.2% (Zizyphus spp. - P. cineraria system); and 10.0% (P. cineraria system) to 33.7% (P. cineraria - A. nilotica system), respectively. Yield reduction under tree canopies for sesamum ranged from 26.3% (P. cineraria -A. nilotica system) to 46.1% (P. cineraria system).

Table 3: Estimates forage production with tree species in arid region of western Rajasthan

Tree species	Forage yield (kg/ha)	Palatable grass (%)
P. cineraria	1500	74.8
Zizyphus sp.	1000	87.5
A. Senegal	600	52.5

Fuel wood and leaf fodder production potential of P. cineraria - A. nilotica system maximum, while it was minimum for Zizyphus spp. - P. cineraria - Salvadora spp. system. Thus, productions from woody components of traditional agroforestry systems follow the rainfall gradient like that of arable crops. (Table 4).

Table 4: Fuel wood and top feed production from traditional cropping systems in arid region

Agroforestry systems	Plant density (Nos./ha)	Fuel wood (t/ha/yr)	Leaf fodder (t/ha/yr)
P. cineraria-A. nilotica based	31.4	1.23	0.29
P. cineraria based	14.2	0.60	0.20
Zizyphus spp. - P. cineraria based	91.7	0.62	0.90
Zizyphus spp. - P. cineraria - Salvadora sp.	17.2	0.18	0.14

Fodder from alternate landuse systems

The Millennium Ecosystem Assessment has categorized the agroforestry ecosystem services into provisioning services (fuel-wood, fodder, timber, poles, etc.), regulating services (hydrological benefits, micro-climatic modifications), supporting services (nutrient cycling, agro-biodiversity conservation), and cultural services (recreation, aesthetics). Agroforestry land use increases livelihood security and reduces vulnerability to climate and environmental change.

Top feed fruit species

The fruit tree species, such as ber, jharberi, boradi, khejri, goonda, pilu, ker, tamarind, etc. are commonly used as top feed by the rural mass for sustaining their livestock in arid region (Table-5). Different clones of thornless type Cactus pear has been introduced from USA, Italy, Mexico, Israel, Argentina, which can be also be used as fodder for animals.

Table 5: Top feed species and their palatability in hot arid region.

Crops		Edible produce	Preferred animals	Palatability ratings
Ber/ Boradi	Ziziphus spp.	Dry leaves i.e. pala	Feed for sheep/goat	Good
	Z.mauritiana Z. rotundifolia	Dry fallen fruits	Feed for sheep	Good
Khejri	P. cineraria	Leaves i.e. loong	Cattle feed	Good
Date palm	Phoenix dactylifera	Seed/stones	Poultry feed	Good
Pilu/ Jaal	Salvadora spp.	leaves	Feed for camel	Moderate
Cactus pear	Opuntia ficus -indica	Stem/cladodes	Fodder	Good
Lasora	Cordia myxa	leaves	Goat, sheep , camel	Moderate
Drumstick	Moringa sp.	leaves	Goat, sheep , camel, cattle	Good
Ker	Capparis deciduas	Tender shoots	Goat	Moderate
Mulberry	Morus spp.	leaves	Feed for sheep/goat	Good

Ber (Ziziphus sp.)

Ber leaves are valuable source of fodder, as it has crude protein 13-17% and fiber 15%. It is popular fodder tree in arid region since it grows and regenerates very fast even under moisture stress conditions. Leaves are also food for silkworms. Dried leaves after pruning of ber serve as good quality forage (Pala) for small ruminants (goat/sheep and other desert animals). Besides, fallen fruits can also be utilized as a source of poultry feed supplement (Pareek, 2001).

In rainfed areas of arid and semi-arid India, kharif and rabi crops are grown successfully with ber plantations. The ber is planted at 6x6 m distance and inter space between the two rows is being utilized to raise the inter-crops. Among various crop species, legumes have an edge over the non-legumes in terms of biological nitrogen fixation. Work conducted at ICAR-Central Arid Zone Research Institute (CAZRI), Jodhpur has revealed that a farmer can get an additional production of 7.82 q/ha cluster bean (Cymopsis tetragonaloba) when grown with jujube plantation in kharif. Under semi-arid conditions of Pali (Rajasthan) 4.36 q ha^{-1} mung bean (Vigna radiata), 10.23 q ha^{-1} clusterbean and 1.88 q ha^{-1} sesame (Sesamum indicum) can be produced under ber (cv. Seb) based cropping system during kharif (Singh and Kumar,1993).

Khejri (Prosopis cineraria)

Khejri plants have long been used by mankind in many ways. In desert, khejri is called, Kalpavriksh' because of its many uses. It has vast socio-economic importance in arid region. Khejri tree is worshipped by a community of people for God of snake i.e. Gogaji in western Rajasthan. Khejri green pods are locally called ''sangari'' or ''sangri''. The dried pods locally called ''Kho-Kha'' are eaten.

Thar Shobha (Khejri Selection-1) was recommended in 2007 (Samadia et al., 2007) and popularized for commercial cultivation. Khejri trees can be successfully propagated by patch budding or bud grafting technique (Pareek, 2001), and therefore, systematic initiatives were taken at ICAR-CIAH from the year 2001. Wild type kheri plants can be converted in to improved type khejri (Pareek, 2002). For this, its selected elite tree, having a record of producing good quality edible pods, were pruned during December to induce new growth bearing juvenile scion buds from March.

Moringaoleifera (Drumstick)

Moringa is a multipurpose tree which grows in all types of soil, from acid to alkaline and at altitudes from sea level to 1800 m. Its roots penetrate deep into soil to search for water and nutrients, which enables Moringa trees to tolerate severe conditions (Ramchandran et al. 1980). Aside from the features of its root system, this species has a fast growing habit, a low requirement of maintenance in later stages, reduced necessity of fertilizers and irrigation, and a high capacity to re-sprout after harvesting.

Cactus pear (Opuntia ficus indica (L.)Mill.)

Cactus pear (Opuntia ficus-indica (L.) Mill) commonly known as Prickly pear or Tuna. It is commercially grown in more than 20 countries of the world for fruits, vegetables, fodder and for preparation of several products Young tender cladodes (Nopalitos) are used as green vegetable in Mexico, Chile, Italy, Israel and South Africa. It is also used as salad. Nopalitos are rich source of carbohydrate (4.5%), fibre (4-6%), protein (1-2%) and minerals, principally calcium (1%).

Mulberry (Morus sp.)

Its leaves and tender shoots have high palatability and left over stalk contains 11.4 % crude protein 3.4 % crude fibre, digestible crude protein content 7.8 % with total digestibility 48.4 per cent. Mulberry is grown as fodder plant for small ruminant in many countries (Srivastava et al., 2006).

Pilu or jaal (Salvadora sp.)

It is used as cattle feed for camel in desert and drought tolerant tree. It grows profusely under saline, salt affected marginal soils. It is a multipurpose tree because of industrial uses of fatty acids obtained from seeds. In Namibia, it is used as a drought-resistant fodder plant for cattle.

Arid vegetables

The green and dry plant parts are used as cattle feed in arid region. In arid region, leguminous and cucurbitaceous crops are mainly grown under rainfed conditions. Under drought conditions, vegetable crops/ waste are used as fodder for animals. Cowpea, moth bean, cluster bean, etc. are used as nutritious fodder. These crops are highly suitable for inter cropping with fruit trees under rainfed conditions.

Agro- Horticulture System

Leguminous crop sown under ber (Zizyphus mauritiana cv. Seb) plantation produced 0.2 t/ha of grain and 0.8 t/ha quality ber fruits from same land unit even when seasonal rainfall is 200 mm, thus rendering a drought proofing mechanism to the system. The density of ber plants were kept 400 individuals/ha. The economics of this improved system indicated that in case of sole leguminous crop (mung bean) farming, the net profit per hectare was Rs. 4800/-, however, in case of ber inter-cropping, the profit was to a tune of Rs. 8000/- per ha.

Table 6: Improved agroforestry practices: agri-horticultural (Ber + Mungbean)

Treatment	Annual rainfall (mm)	Fruit yield (kg/ha)	Grain yield (kg/ha)	Net profit (Rs/ha)
Sole crop	200	-	520	4800
Inter-cropped *with Ber*	200	800	200	8000

Horti-Pasture system

Ber based horti-pasture have proved highly remunerative in Thar desert on farmers' field. Ber trees were planted at spacement of 6x6 m and grass Cenchrus ciliaris was introduced between tree rows after third year. On an average, the dry grass production was 1.55t/ha/year (Tewari et al., 1999) (Table 7). The fruit, leaf fodder and fuel wood production from ber was 2.77, 1.87, 2.64 t/ha/year, respectively.

Table 7: Improved hortipasture (Ber + Cenchrus ciliaris) on farmers' field

Year	Dry grass (t/ha)	Tree products		
		Fuelwood (t/ha)	Leaf fodder (t/ha)	Fruit yiel d (t/ha)
1	1.50	3.10	1	1.50
2	1.67	2.74	2	1.67
3	1.42	2.46	3	1.42
4	1.66	2.49	4	1.66
5	1.48	2.43	5	1.48
6	1.57	2.63	6	1.57
Average	1.55	2.64	Average	1.55

Selection of grass species

The improved grasses viz. Anjan (Cenchrusciliaris), Dhaman (Cenchrussetigerus) and Sewan (Lasiurussindicus) can be taken in horti-pasture system in dry areas. Such system fulfillsthe fodder need of the farmer along with cash returns in the form of fruits (Table 5).Cenchrusciliaris (dahman) in Jodhpur and Nagaur, and Lasiurussindicus (sewan) in Jaisalmer, Barmer, and Bikaner, while Cenchrussetigerus (dahman) is preferred in Pali, and in areas having a MAR in excess of 350mm in general.

Table 8: Suitable grasses for western Plains, hot arid with desert and saline soils

Particulars	Grass species
sandy plains	*Cenchrusciliaris, Cenchrussetigerus*
sand dunes	*Dichrostachyscinerea,Panicumturgidum*
salt-affected land	*Chlorisgayana, Sporobolusmarginatus*

1. Cenchrus ciliaris (Anjan)

Anjan grass is also commonly known as buffel grass and African foxtail grass. It is a good quality palatable grass for livestock which is a tall, thick stemmed, erect and drought hardy perennial. Its leaves are broad, long droopy and remain green upto maturity. Seeds remain viable for 2 - 3 years. In India, it is a natural species largely found in Rajasthan, Haryana, Punjab, Gujarat and parts of western Uttar Pradesh and Tamil Nadu state. It grows well in rainfall zones ranging between 125 and 1,250 mm in arid and semi-arid regions of the country. In the first year of establishment, only one cut is to be taken in mid-October. From first year onward, this grass gives 2-3 cuts.

2. Cenchrus setigerus (Dhaman)

Cenchrus setigerus, commonly known as Birdwood grass, is one of the most drought-tolerant perennial grasses. Native to East Africa and Western Asia between latitudes 30p N and 30p S, this grass occurs naturally in areas with average rainfall between 125 mm and 1250 mm. The tufted grass is very resilient, withstanding low and sporadic rainfall, long dry seasons, strong winds, high temperatures, frost, erosion, nutrient-poor soils, and salinity. Cenchrus setigerus is well adopted in the arid and semi-arid regions of India though require little higher soil moisture than Cenchrus ciliaris. In Rajasthan, growth peaks from August to October. The grass flourishes in well-drained sandy loams and, although it tolerates arange of other soils from alkaline to heavy clay, it rarely grows in stony clay soils or near water courses. Once established, the grass is very hardy. Hardiness and strong roots mean that the grass responds quickly to rain even though it may appear to be dead. Freshly harvested Cenchrus setigerus seed has a germination rate of 10–20%. The seed fuzz contains toxic inhibitors. Soaking the seed for 8–12 hours before sowing removes this toxic fuzz.

3. Lasiurus sindicus (Sewan grass)

Lasiurus sindicus (Sewan) is the primary grass of extremely arid parts of Jaisalmer, Barmer and Bikaner districts of western Rajasthan in the Indian Thar Desert. Lasiurus sindicus is well grown in the 200-mm mean annual rainfall of arid region. It thrives well under moisture stress on sandy plains, low dunes and hummocks of this region, receiving annual rainfall below 200 mm. Until the last decade, about 80 per cent of the total geographical area of Jaisalmer covering Nachana, West Puggal,Mohangarh, Sultana and Binjewala with 100–150 mm annual rainfall supported Sewan grasslands. Fordecades, these grasslands in turn supported an ever increasing livestock population in the Indian desert (Shankar and Kumar, 1987).

Forage conservation and utilization

In most of the developing countries feed shortage is one of the major constraints in improving livestock production and India with a large human population and limited land mass is no exception. At present, the country faces a net deficit of 61.15% of green fodder, 21.9% dry fodder crop residue and 64% concentrate feed. 54% of the total fodder is met from the crop residue, while 18% fodder is met from grassland and only 28% fodder is met from cultivated fodder crops (Hedge, 2006).

Forages and livestock have been the integral component of farming system in India for ages. Forage may be defined as the vegetative matter, fresh or preserved, utilized as feed for animals. Forage crops include grasses, legumes, crucifers and other crops cultivated and used in the form of hay, pasture, fodder and silage. Fodder crops are the plant species that are cultivated and harvested for feeding the animal in the form of forage (cut green and fed fresh), silage (preserved under anaerobic conditions) and hay (dehydrated green fodder).

Need of conservation of forage crops

Availability of nutritious forage throughout the year is very essential for profitable livestock farming. But in country like India, it varies from season to season. Therefore, every dairy farm must preserve the surplus forages in the form of silage or hay. The surplus forages of the two glut seasons i.e. rainy season (August-September) and winter season (February-mid March) could easily be carried over to the succeeding lean periods of fodder supply. During the lean periods of May and mid July and November-December, the hay or silage can supplement the dry fodder and limited quantity of green fodder available to overcome the scarcity of fodder.

Methods for conservation of Forage Crops : There are two methods of conserving forages:

i) **Hay** - the simple method is to drive off moisture in forages.

ii) **Silage** - natural fermentation is facilitated to retain succulence in the preserved forage. The driving off moisture from forage forms the basis in hay making while retaing forage's succulence forms the basis for silage making.

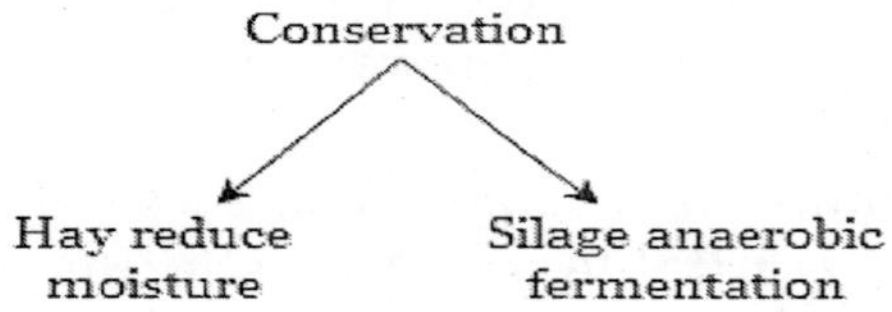

Hay : Reducing the moisture content of the green crop to a level low enough (12-15%) to inhibit the action of plant and microbial enzymes is the aim of hay making. The harvested crop can be dried either by natural drying or through artificial drying, but natural drying is preferred, as there it can be done without incurring expenditure towards electricity. Hay can be stored satisfactorily in a stack or bale.

Requisites for good hay : Good quality hay can be produced by considering the following points:

Selection of crop: The crop to be made as hay should have soft pliable stem.

Harvesting of crop: The crop should be harvested at 2/3rd flowering stage as it is at that time the plant will have the maximum nutrient in it. Delaying the harvesting further would divert the nutrients from the plant to seed production resulting in low nutritive value of the harvested crop. Hay should be leafy and green in colour as they reflect the nutritive value of hay. Hay should be free from moulds and weeds. Hay should have the characteristic aroma of the crop.

Harvesting of crop and curing of hay : Good quality hay can be produced by harvesting the crop early in the morning and left in the field as such for curing. The harvested crop should be allowed to dry in the field until the moisture content is reduced to about 40%. Frequent turning is necessary to facilitate uniform drying. On sunny days field drying of harvested crop for two days is sufficient to make hay. The air dried crop may be turned with the rake and made into small feathery windrows at the end of first day.

Management systems and suggested interventions

Historically, management of common lands followed at least three different patterns. In some places, they were accessible to all, with insufficient pressure on resources to lead to severe degradation until the last several decades. In others, powerful landowners, such as zamindars, who acted as "gatekeepers" enforced management to make sure that the common lands were not overexploited (Gadgil and Guha 1992). While this system was good for the condition of the land, it was inequitable, with benefits dominated by the landlords. In a third kind of situation, democratic village-level institutions resulted in sophisticated, equitable ways of sharing both rights and responsibilities for managing common lands (Agarwal and Narain, 1989). Management systems for utilizing common grazing lands include uncontrolled grazing (i.e. passive management), privatisation, and special projects that aim to raise productivity while also restricting access through social institutions. In addition, with the decline in area and productivity of common lands people have looked to other areas for supplying fodder or earning their livelihoods.Common pasture management is facing the following constraints:

- Encroachment (more serious than for forest)
- Long-term security of use rights (3-5 years leases from PRI)
- Conflicts between and within villages (heterogeneity)
- Indifferent attitude of revenue department
- Benefits sharing arrangements
- High livestock population to large, and
- Lack of long-term plans and analysis in relation to sustainability

There is a large gap between policies and programmes to tackle the issue of dependence on common lands for fodder (Anonymous, 2011). Although stall feeding of productive livestock and producing fodder in high yielding agricultural areas is recommended, the economics of raising fodder in drier areas or of the relative merits of stall feeding compared to regulated grazing is not well understood. High yielding cattle breeds cannot be supported by rural poor through stall feeding in most areas of India and even less so in dry regions. Yet programmes promoting livestock development continue, as though there were no conflicts with wasteland and forest rehabilitation programmes.

Way Forward

There is abundant evidence of the declining productivity of common grazing areas in arid and semi-arid India from the change in the composition of plant species, the widespread change in the composition of livestock from cattle to sheep and goats, and the reduction in the effective period of sustained grazing. The response of farmers to the decline in grazing land productivity has been three-fold: (i) they attempt to maximize the complementarities between common and private property resources; (ii) they accept inferior options in common grazing areas because of a lack of alternatives; and (iii) they resort to measures of desperation, such as increased frequency and unseasonal grazing, leading to reduction in seed formation, removal of plant roots and the use of hitherto inferior grasses and legumes, with corresponding effects on livestock productivity. Currently, many watershed development projects aim to revive the productivity of common grazing lands. (Farrington et al, 1999; Kerr et al, 2002). They aim to use a combination of technical and institutional means to move the supply of products, such as fuel and fodder from a low-level equilibrium to a high-level one. The technical work involves installing soil and water conservation works and planting vegetation to raise pasture productivity, and also transforming herds from mainly relatively unproductive grazing animals to mainly high productive stall-fed animals (NRAA, 2009). In addition, many projects also seek to develop institutions for managing government lands based on principles of common property resource management. Typically, they encourage villagers to establish users' committees that are expected to develop and enforce management plans in a way that satisfies the needs of every interest group. In short, they try to create the kind of ideal, democratic arrangement mentioned above. Although this approach has succeeded in a small number of cases, it remains too soon to know if it can do so on a large scale.

References

1. Birbal, Subbulakshmi V., Sheetal KR., Renjith PS. and Soni M.L. 2016. Horti-Pasture System - A Boon for Dry land Dwellers. Lecture note in training programme on Forage Resource Management for Sustainable and Economic Livestock Production from Nov. 29 – Dec. 08, 2016 at DHRD, SKRAU, Bikaner (Raj.) India.

2. Jodha, N.S. 1996. Common property resources and dynamics of rural poverty in India's dry regions, Unasylva, An international journal of forestry and forest industries, United Nations Food and Agriculture Organisation, Vol. 46.

3. Mishra A K and R N Kumawat.2016. Range management: Traditional knowledge and technological interventions to cope with uncertainties. Lecture note in training programme on Forage Resource Management for Sustainable and Economic Livestock Production from Nov. 29 – Dec. 08, 2016 at DHRD, SKRAU, Bikaner (Raj.) India.

4. NRAA 2009. Harnessing opportunities in rainfed area: A pathway to prosperity vision 2025. National Rainfed Area Authority, New Delhi.

5. Planning Commission 2011. Report of the Sub-Group III on Fodder and Pasture Management Constituted under the Working Group on Forestry and Sustainable Natural Resource Management, Planning Commission, GoI, Version:1.5 (21 September, 2011).

6. Pareek,O.P.2001. .Ber, International Centre for Underutilized crops Southampton,U.K., p. 290.

7. Pathak, P.S. and Sunil Kumar. 2004. Forage and grazing resources in different agro-climatic regions in India In : S.S. Kundu, A.K. Mishra and P.S. Pathak (eds.). Buffalo production under different climatic regions. International Book Distribution Co., Lucknow p. 1-41.

8. Rajesh Nehra.2016.Forage conservation and utilization. Lecture note in training programme on Forage Resource Management for Sustainable and Economic Livestock Production fromNov. 29 – Dec. 08, 2016 at DHRD, SKRAU, Bikaner (Raj.), India.

9. Samadia, D. K. 2007. Sangri utpadan kai leai khejri ki Bagwani. Technical bulletin (Hindi) - CIAH/ Tech/Pub. No., 20, pp 1-32.

10. Saroj, P.L. Dhandar, D.G. and Singh ,R.S. 2004. Indian Aloe, Tech. Bull. CIAH, Bikaner

11. Saroj, P.L. R.S. Singh and Hare Krishna.2016.Forage Production from Horticultural Crops for Livestock feed under Hot arid regions.

12. Singh, Gurbachan, 1995. An agroforestry practice for the development of salt lands using Prosopisjuliflora and Leptochloafusca. Agroforestry Systems 27: 61-75.

13. Sunil Kumar, D. R. Palsaniya, Mukesh Choudhary and S. R. Kantwa, 2016. Climate Resilient Forage Production Strategies for Arid and Semi-arid Regions. Lecture note in training programme on Forage Resource Management for Sustainable and Economic Livestock Production from Nov. 29 - Dec. 08, 2016 at DHRD, SKRAU, Bikaner (Raj.), India.

❑❑❑

Chapter 19

Livestock Based Enterprise for Doubling the Farmer's Income

Dr. C.M. Yadav

Scientist (Animal Production), KVK, Bhilwara

Introduction

Livestock plays an important role in Indian economy. About 20.5 million people depend upon livestock for their livelihood. Livestock contributed 16% to the income of small farm households as against an average of 14% for all rural households. Livestock provides livelihood to two-third of rural community. It also provides employment to about 8.8 % of the population in India. India has vast livestock resources. Livestock sector contributes 4.11% GDP and 25.6% of total Agriculture GDP.

India is world's highest livestock population at about 512.3 million. First in the total buffalo population in the world-105.3 million buffaloes. Second in the population of cattle and goats-140.5 million goats. Second largest poultry market in the world-production of 63 billion eggs and 649 million poultry meat. Third in the population of sheep (72 millions). Fifth in the population of ducks and chicken. Tenth in camel population in the world.

Table 1: Livestock population (2012 Livestock census)

Species	Number (in millions)	Ranking in the world population
Cattle	190.9	Second
Buffaloes	108.7	First
Total (including Mithun and Yak)	300	First
Sheep	65.0	Third
Goats	135.2	Second

Pigs	10.3	-
Others	1.7	-
Total livestock	512.3	
Total poultry	729.2	Seventh
Duck	-	Fifth
Chicken	-	
Camel	-	Tenth

Table 2: Production of livestock in India

Product	Quantity	Ranking in the world production
MILK in million tonnes	137.7	First
EGGS in billions (nos)	74.75	Third
MEAT million tonnes	8.89	Na
WOOL in million kgs.	47.9	Na
FISH in lakh tonnes	95.79	Second

Table 3: Productivity of livestock (Census – 2012)

Category	Population	Productivity
Cow	7,19,472	
Crossbreed	73,262	2.380 liter Milk / day
Indigenous	6,46,210	
Buffalo	3,82,056	4.203 liter Milk / day
Sheep	5,37,839	
Crossbred	4,795	-
Indigenous	5,33,013	-
Goats	8,73,218	0.465 liter Milk / day
Poultry	82,480	-
Indigenous	62,081	50-60 egg/year
Improved (Pratapdhan)	20,339	150-180 egg/year

Contribution of livestock to people

The livestock provides food and non-food items to the people:

1. **Food :** The livestock provides food items, such as Milk, Meat and Eggs for human consumption. India is number one milk producer in the world. It is producing about 163.7 million tones of milk in

a year (2016-17). Similarly, it is producing about 88.81 billions of eggs, 8.89 million tonnes of meat in a year. The value of output of livestock sector at current prices was Rs. 5,91,691 crore in 2015-16.

2. **Fibre and skins :** The livestock also contributes to the production of wool, hair, hides, and pelts. Leather is the most important product which has a very high export potential. India is producing about 47.9 million Kg of wool per annum.

3. **Draft :** Bullocks are the back bone of Indian agriculture. Despite lot of advancements in the use of mechanical power in Indian agricultural operations, the Indian farmer especially in rural areas still depend upon bullocks for various agricultural operations. The bullocks are saving a lot on fuel which is a necessary input for using mechanical power like tractors, combine harvesters, etc. Pack animals like camels, horses, donkeys, ponies, mules, etc. are being extensively used to transport goods in different parts of the country in addition to bullocks. In situations like hilly terrains mules and ponies serve as the only alternative to transport goods. Similarly, the army has to depend upon these animals to transport various items in high areas of high altitude.

4. **Dung and other animal waste materials :** Dung and other animal wastes serve as very good farm yard manure and the value of it is worth several crores of rupees. In addition, it is also used as fuel (bio gas, dung cakes), and for construction as poor man's cement (dung).

5. **Storage :** Livestock are considered as "moving banks" because of their potentiality to dispose off during emergencies. They serve as capital and in cases of landless agricultural labourers, many time it is the only capital resource they possess. Livestock serve as an asset and in case of emergencies they serve as guarantee for availing loans from the local sources, such as money lenders in the villages.

6. **Weed control :** Livestock are also used as Biological control of brush, plants and weeds.

7. **Cultural :** Livestock offer security to the owners and also add to their self esteem especially when they are owning prized animals, such as pedigreed bulls, dogs and high yielding cows/ buffaloes, etc.

8. **Sports / recreation :** People also use the animals like cocks, rams, bulls, etc. for competition and sports. Despite ban on these animal competitions the cock fights, ram fights and bull fights (jalli kattu) are quite common during festive seasons.
9. **Companion animals :** Dogs are known for their faithfulness and are being used as companions since time immemorial. When the nuclear families are increasing in number and the old parents are forced to lead solitary life the dogs, cats are providing the needed company to the latter thus making them lead a comfortable life.

Role of livestock in farmer's economy

The livestock plays an important role in the economy of farmers. The farmers in India maintain mixed farming system i.e. a combination of crop and livestock, where the output of one enterprise becomes the input of another enterprise thereby realize the resource efficiency. The livestock serve the farmers in different ways:

1. **Income :** Livestock is a source of subsidiary income for many families in India, especially the resource poor who maintain few heads of animals. Cows and buffaloes, if in milk will provide regular income to the livestock farmers through sale of milk. Animals like sheep and goat serve as sources of income during emergencies to meet exigencies like marriages, treatment of sick persons, children education, repair of houses, etc. The animals also serve as moving banks and assets which provide economic security to the owners.
2. **Employment :** A large number of people in India being less literate and unskilled depend upon agriculture for their livelihoods. But agriculture being seasonal in nature could provide employment for a maximum of 180 days in a year. The landless and less land people depend upon livestock for utilizing their labour during lean agricultural season.
3. **Food :** The livestock products, such as milk, meat and eggs are an important source of animal protein to the members of the livestock owners.
4. **Social security :** The animals offer social security to the owners in terms of their status in the society. The families, especially the landless which own animals are better placed than those who

do not. Gifting of animals during marriages is a very common phenomenon in different parts of the country. Rearing of animals is a part of the Indian culture. Animals are used for various socio religious functions. Cows for house warming ceremonies; rams, bucks and chicken for sacrifice during festive seasons; Bulls and Cows are worshipped during various religious functions. Many owners develop attachment to their animals.

5. **Draft :** The bullocks are the backbone of Indian agriculture. The farmers, especially the marginal and small, depend upon bullocks for ploughing, carting and transport of both inputs and outputs.

6. **Dung :** In rural areas, dung is used for several purposes which include fuel (dung cakes), fertilizer (farm yard manure), and plastering material (poor man's cement).

Scope of Livestock Based Enterprises

Livestock based enterprises improve productivity, profitability, sustainability, balanced food, environmental safety, recycling, income rounds the year, saving energy, meeting fodder crisis and employment generation.

Live Units

1. Poultry Unit

It was established in 2013 with a poultry breed- "Pratapdhan". Currently 14 (Birds) are there in aforesaid unit. It has Net Income of around Rs. 4000/month/20 Birds.

Pratapdhan : dual purpose coloured bird for Rajasthan

Pratapdhan is a dual purpose chicken variety to cater to the needs of rural poultry keepers of Rajasthan. It was developed as part of AICRP on Poultry Breeding by MPUAT, Udaipur. It resembles local birds of Rajasthan. Attractive multicolour feather pattern, as rural people like coloured birds from aesthetic point of view and better looking. Because of colour plumage, birds have camouflagic characters to protect themselves from predators. Birds have longer shank length which helps in self-protection from predators in backyard areas and has capacity to survive on low plane of nutrition (low and negligible input) and harsh climatic conditions. It lays brown eggs weighing around 50 g and has broody characteristic to some extent. It has fast growth rate with average adult body weight at 20 weeks of age ranged from 1478 to 3020 g in males and 1283 to 2736 g in females. The age at sexual maturity was 170 days. Pratapdhan produces 161 eggs annually, which is 274% higher than local native (43 eggs).

2. Sheep Unit

Sheep Unit Established in 2014 with a breed- "Kendrapara". Currently it has 5 sheeps. It has Net Income of about (Rs.) 1500-2000/ Year/Head.

About Kendrapara Sheep

- Kendrapara sheep is found only in coastal Jagatsinghpur and Kendrapara districts of Odisha. Locally it is called kuji mendha.
- The average adult sheep weighs 18-20 kg and are dwarf in built with the body covered with coarse hair. They are well adapted to high ambient temperature, high humidity and heavy rain.
- Kendrapara sheep carries FecB mutation gene, which is responsible for prolificacy (multiple babies in same delivery) or multiple birth syndrome.
- This characteristic makes them distinctive from other species sheep species, as they are not known for giving multiple births.
- Kendrapara sheep is second sheep breed in India (other one is Garole sheep found in West Bengal) and the sixth in the world that carries FecB mutation gene.

- Economic Significance : Multiple-birth characteristics make it a profitable livelihood source. They are primarily used for mutton production. Besides, their skin also has economic importance.

3. Goat Unit

Goat unit was established in October, 2012 with a breed "Sirohi". Presently they have 11 Goats. It has Net Income of around (Rs.) 4000/ Year/Head.

The Sirohi goat is an Indian breed, which is native to the Sirohi district of Rajasthan. The name ***Sirohi*** derived from the home tract. The breed is also known as Ajmeri, Devgarhi and Parbatsar. Sirohi goats can be found over the Ajmer, Bhilwara, Jaipur and Tonk districts. Usually Sirohi goats are raised under extensive grazing system in commercial farms, and most of the flocks are kept confined. Some rural farmers used to migrate their large sized flocks seasonally to nearby areas for the search of foods and water. The flock size varies from place to place. And a flock of 10-200 Sirohi goats can be found in different places of Rajasthan and Uttar Pradesh. But today, the Sirohi goat is very popular as a meat and milk goat breed in India. Now the goat breed is found in almost every state of India.

Sirohi goat is a medium-sized animal. It has pre-dominantly brown coloured coat with dark or light brown patches. On occasion, the colour of a very few goats may be completely white. The body is covered with fairly dense soft and coarse hair. The hair grows at the rate of about 2 centimetre per year. Sirohi goats have medium-sized flat dropping ears (much like leaf), and most of the goats are wattled. Both Sirohi buck and doe have small horns which is curved upward and backward. Their tail is curved upward and medium in length. Their legs are long and very strong. The udder of Sirohi doe is small and round, with small teats placed laterally. Sirohi bucks are much heavier than the does. And the average body weight of an adult buck is upto 50 kg and doe is upto 30 kg. Sirohi goat is a meat producing breed, and mainly raised for meat production purpose rather than milk.

4. Dairy Unit

Dairy unit was established in 2015 which has a breed "Gir". Currently position 4+5 (Calf). It has Net Income–(Rs.) 1500-2000/ month/Head.

The Gir cattle is a famous dairy cattle breed originating in India. It is one of the principal ***Zebu*** or ***Bos indicus*** breeds in India, and has been used locally in the improvement of other breeds. The native tract of Gir cattle is Gir hills and forests of Kathiawar, which includes Amreli, Bhavnagar, Junagadh and Rajkot districts of Gujarat. The breed derives its name from the Gir forest, which is the natural habitat of the breed. The breed is also known as some other names in different parts of their breeding tract. Such as ***Bhodali, Gujarati, Sorthi, Surti, Kathiawari*** and ***Desan***. Gir cattle is also very popular outside India. United States, Mexico, Venezuela and Brazil have imported this breed. And they are being bred successfully there. The breed was also one of the cattle breeds used in the development of the Brahman cattle in North America.

Gir cattle is distinctive in appearance and a heavy breed. Their body is well proportioned and the udder in cows is well developed and round. Their body colour is shining red to spotted white. Although, their body colour can vary from pure red to speckles, yellowish red to white with large red spots. Their skin is soft, thin and glossy. They have a big head with prominent bulging sheild and a long face. The convex forehead of Gir cattle is the most unusual feature of this breed. Their convex forehead acts as a cooling radiator to the brain and pituitary gland. Their ears are very long and are an excellent fly and insect swatter. Their eyes are hooded and black pigmented and there are lot of loose skin around their eye area. They have very hard feet which is black in colour. The horns of Gir cattle are set well back on their heads and thick at the base. Their horns grow downwards and backwards with an upward curve. They have large hump. The average body weight of Gir bull is about 545 kg and 385 kg for cows. Gir cattle are a dairy cattle breed and mainly raised for milk production in India. This breed has also been used locally in the improvement of some other breeds.

5. Rabbit Unit

Rabbit unit was established in January, 2015 which has breed "Grey Giant". Present position of rabbit unit is 20. It has Net Income – (Rs.) 2000/Year/Head

6. Duck & Fish Unit

Duck and Fish unit was established in August 2017. It has breed of "White Picken". Presently I has position 2 + 2.

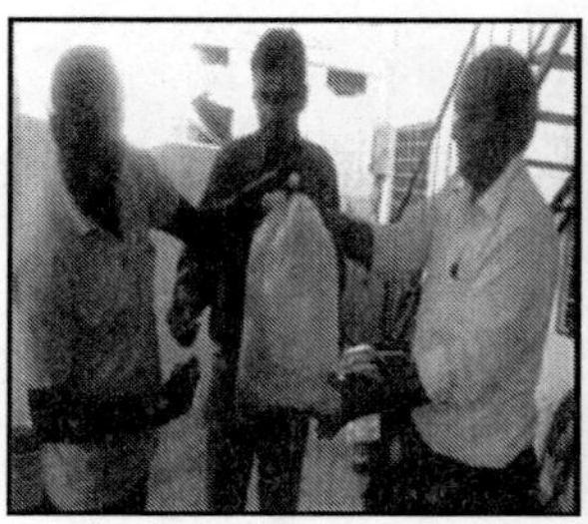 

7. Quail Unit

It was established in December, 2017. It has breed of "Japani Quail". It has currently 6 + 4. It has Net Income of around Rs. 4500/ month/20 Quail.

Quail farming is very profitable like other farming ventures, such as chicken, turkey or duck farming business. Almost all types of weather conditions are suitable for starting quail farming business. Meat and eggs of quail are very tasty and nutritious. Quail eggs are very nutritious than other poultry eggs. Because quail eggs contain comparatively more protein, phosphorus, iron, vitamin A, B1 and B2. Quail farming can play a vital role to meet up the demand of food and nutrition. Quail farming business require small capital and labour. Quails can be raised along with your other poultry birds for meat or eggs production.

Characteristics of Quail

- Quails are very small-sized bird.
- An adult quail weights between 150 to 200 grams and an egg weights around 7 to 15 grams.
- Female quails start laying eggs within their 6 to 7 weeks of age and continuously lay one egg daily.
- They lay about 300 eggs in their first year of life. After that they produce about 150 to 175 eggs in second year. Eggs productions gradually decrease after their first year of laying period.
- Quail egg is very suitable for human health. It contains 2.47% less fat than chicken egg. Many people believe that 'quail eggs help to prevent blood pressure, diabetic, pant, etc'.
- Quail meat is very tasty and nutritious. Fat is very low in their meat. So, quail meat is very suitable for blood pressure patients.
- Eggs are very beautiful with multiple colour.
- Quails do not incubate their eggs. So, you have to use an incubator or brooder chickens for hatching their eggs.

Benefits of Quail Farming

Quails are smaller-sized poultry birds, and it's very easy to handle them. The main benefits of starting quail farming business are listed below:

- Quails are smaller-sized bird, so they can be raised within small place.
- Feeding cost of quails are comparatively lower than chickens or other poultry birds.
- Diseases are less in quails, and they are very hardy.
- Quails grow very fast and gain maturity faster than any other poultry birds.

- They start laying eggs within their 6 to 7 weeks of age.
- It takes about 16 to 18 days to hatch their eggs.
- Meat and eggs of quail are very tasty, delicious and nutritious. So, it's a great source of food and nutrition.
- Quail farming needs small capital, and labour cost is very low.
- Quails can be raised successfully in commercial method. Some people have already started commercial quail farming business.
- Quails are very strong bird and diseases or other health problems are less. So, risks are less in this business.
- Quail meat contains less fat. So, it is suitable for high blood pressure patients.
- Their food to meat or eggs converting efficiency is satisfactory. They can produce one kg meat or eggs by consuming three kg food.
- You can raise 6 to 7 quails in the same place that is required for one chicken.
- As the quail eggs are smaller in size, so the price is also lower than other bird's egg. As a result, all types of people can buy quail eggs and you can easily sell the eggs.
- As the primary costs are less, so you can start this business with a very little investment.
- You can raise about 6 to 8 quails within 0.91 squire metre area.
- As it is a lucrative business venture, so commercial quail farming business can be a great source of income and employment for the unemployed educated people. Even, you can start raising a few quails along with your current profession.

8. Vermicompost Unit

It was established in 2014-15 with an area of 18X60Ft. It has Net Income (Rs.) 30000/Year.**Vermicomposting** is the usage of earthworms to convert vegetable waste to a 100% natural plant fertilizer. Vermiculture is the side issue of the breeding of common earthworms for use in vermicomposting. The use of worm farms for vermicomposting is becoming a favourite way of converting waste to a valuable product while also growing more worms to increase the capacity of the worm farms. Advantages of vermicompost include enriching soil, increasing harvest yields and suppressing plant disease.

9. Vermiwash Unit

It was established on November, 2015. It reduces fertilizer dose and increase productivity of crop & vegetable. **Vermiwash** is a brown coloured liquid fertilizer, which is collected after water passes via a worm culture column. As storehouse of nutrients and micro-organisms, **Vermiwash** is used as a foliar spray for crops.

10. Low Cost Hydroponic Unit

It was established in December, 2016 with a cost of 1.50 Rs./kg

- Superior nutrition and hydration
- No fertilizer, herbicides or pesticides
- Low water consumption; ideal for drought areas
- Most economically feasible for grass-fed milk production, organic-certified operations, and areas where sufficient and consistent pasture production is limited, hay is scarce or expensive, and for drought-prone areas.

11. Fodder Park

Availability of green fodder round the year .Feeding of fodder will also provide these advantages to cattle:

- Increased digestibility

- Earlier breeding and higher conception rates
- Faster weight
- Increased longevity in dairy cows
- Higher milk production

12. Napier Grass

Availability of green fodder round the year **through Napier grass** is an improved fodder **grass** that produces a lot of high-protein forage. It is also known as "**elephant grass**", "Sudan **grass**" or "king**grass**". Its scientific name is Pennisetum purpureum.

- Napier grass is propagated easily.
- It has a soft stem that is easy to cut.
- It has deep root, so is fairly drought-resistant.
- The tender, young leaves and stems are very palatable for livestock.
- Napier grass grows very fast.
- This is a perennial grass fodder.

13. Azolla Unit

It was established on 2012.

The Krishi Vigyan Kendra is popularizing the cultivation of Azolla for supplementing the fodder requirements of the livestock farmers in the district. Azolla is a floating fern mostly utilized as Bio-fertilizer for wetland paddy. It belongs to the family of azollaceace. The fern Azolla, hosts a symbiotic blue green algae anabaena azolle, which is responsible for the fixation and assimilation of atmospheric nitrogen. Azolla was found to be a very nutritive and cheap organic feed substitute for dairy cattle, poultry, pig, fish, etc. Azolla is rich in protein, almost 25-30 % on dry weight basis, it is also found to contain essential minerals like Iron, Calcium, Magnesium, Phosphorus, Copper, Manganese, etc. apart from appreciable quantities of vitamins A and vitamins B_{12}. It is also found to contain almost all the essential Amino acids, many probiotics, Biopolymers and B-carotene. The above mention bio-chemical constitution along with the rapid multiplication rate makes Azolla ideal organic feed substitutes for livestock. Livestock can easily digest Azolla due to high protein content and low lignin content.

Azolla has enormous potential as a livestock feed due to:

- Its high content in proteins, essential amino acids, vitamins (vitamin A, vitamin B_{12}, Beta Carotene), growth promoter intermediaries and minerals.
- Its ability to proliferate without inorganic nitrogen fertilization.
- Its high rate of growth in water without the need to displace existing crops or natural ecological systems.

Azolla meal contains 25.78% crude protein, 15.71% crude fiber, 3.47% ether extract, 15.76% ash and 30.08% nitrogen free extract on the air-dry basis (Basak *et al.*, 2002). In addition, aquatic plant species including *Azolla* do not to accumulate secondary plant compounds and therefore, has a greater potential than tree leaves to source protein for monogastric animals.

Farmer's Feedback

Farmers have positive response towards livestock based enterprises as they found increase in milk production, increase fat% in milk, increase protein% in milk, improve health status of animals, improve economic status of farmers, improve fertility in animals, increase egg production in poultry, increase body weight in broiler and source of employment for farmers.

Conclusion

It may be concluded the Agricultural systems that integrate livestock help the poor small farmers, who have very small land holding for crop production and a few heads of livestock to diversify farm production, increase cash income, improve quality and quantity of food produced and exploitation of unutilized resources.

The livestock enterprises significantly increase the farmer's income by using different models. Azolla as feed substitute shows that there is an overall increase of milk by 10-15% by supplementing feed with the same quantity of Azolla on dry weight basis without affecting the milk production.

□□□

Chapter 20

Land Resource Characterization and Agricultural Landuse Planning towards Enhancing Agricultural Productivity

Ram Sakal Singh
Regional Director, NBSS & LUP, Udaipur

Introduction

Land evaluation is the process of estimating the potential of land for alternative kinds of use. These uses can be productive, such as i) arable farming, ii) livestock production, iii) forestry or other uses, such as, a) catchment protection, b) recreation, c) tourism, d) wildlife conservation. It involves interpretation of surveys, climate, soils, and vegetation and other aspects of land with the requirements of alternative landuse.

Objective of land evaluation

To put the necessary information about land resources that is necessary for planning, development and management decisions: before planner, user, farmer, Govt. official and politician.

End results

Number of clear recommendations, alternatives, appropriate types of landuse together with their consequences.

A multi-disciplinary approach

Soil survey includes Agro-climatology, Geomorphology, Soil conservation, Agronomy, Landuse, Farming systems, Economics and

Sociology. Groups include Geomorphology with soil survey and Sociology with economics.

Landuse planning

Landuse planning is the systematic assessment of land potential, social and economic conditions and alternate patterns of landuse, which are beneficial to landusers without degrading the natural resources. The landuse planning is a multi-disciplinary holistic approach that not only meets the function of the land but also actively involves all the stakeholders through a participatory approach.

Table 1: Data and information needed for land use planning

Land resources data	-Climate -Landforms and soils -Land cover -Water resources
Land use related data	-Present landuse and characteristics -Selected physiological characteristics of crops (as determining ecological requirements) -Land utilization types (LUTs) and production systems (present and potential) -Ecological requirements of LUTs, production systems, landuse
Social-economic data	-Population (including age and gender distribution, stakeholder) -Living conditions (including workload, cultural aspects, traditions, etc.) -Access to markets -Costs of production and product prices -Socio-economics of communities
Legal data and information	-Relevant government policy documents, laws and regulations related to land -Present system of land allocation -Land tenure information -Traditional ownership and user rights
Institutional information	-Involved institutions and their mandates, resources and infrastructure links between institutions support services (extension, etc.)
General data and information	-Infrastructure, accessibility

Factors Influencing Alternate Landuse Options

Climatic Factors : It includes Temperature, Rainfall, Sunshine hours, Relative humidity, Wind velocity and direction, Seasons, Climate Change and Agro-ecological situations.

Soil Characteristics : It includes Soil Depth, Soil texture, Available water capacity, pH, Soil fertility, Salinity & sodicity.

Water : It includes adequate water source like tanks, wells, dams, etc. Suitability of quality irrigation water. Availability of electricity for lifting the water. Availability of pump sets, micro-irrigation systems.

Cropping System Options : Do we have the opportunity to go for inter-cropping, mixed cropping, multi-storeyed cropping, relay cropping, etc. Knowledge on cropping system management.

Past & Present Experience of Farmers : What were your previous experiences with regard landuse options?

Expert opinion on the proposed Landuse options

Expected Profit & Risks : Expected profit from the proposed cropping system. Profit is better than the existing cropping system. What are the risks you are anticipating in the proposed cropping system and solution to manage the risk. Is it worth to take risks for anticipated profits?

Economic condition of Farmers including Land holdings : Are the proposed cropping systems suitable for the size of land holding? Are financial resources adequate to manage the proposed options? If not, can it be mobilized through alternative routes?

Labour Availability & Mechanization Potentials : Availability of labour (Family/hired) to manage proposed cropping system. Are there any mechanization options to substitute labour? Affordability (cost effective) of available machinery/custom hiring. Is family/hired labour equipped to handle the machinery?

Technology Availability and Suitability : Is the proposed cropping system suitable? Availability of technologies for proposed system. Access to extension to get the technologies. Are technologies economically feasible and technically viable? Are technologies complex or user friendly?

Market Demand & Availability of Market Infrastructure : Are the crops proposed in market demand? Availability of market infrastructure to sell produce. Availability of organized marketing system to reduce intermediaries. Availability of real time market information and market intelligence.

Policies and Schemes : Do Government policies favour your crops? Is there any existing scheme which give incentives? Eligibility to avail those benefits.

Public & Private Extension Influence : Do farmers have access to the ATMA, KVK, SAU, ICAR institutions, Agri-business, NGOs, Agriclinics, etc. to get advisory?

Availability of Agricultural Input including Credit : Availability of adequate agricultural inputs, such as seed, fertilizers, pesticides and implements in time. Access to institutional credit.

Post Harvest Storage & Processing Technologies : Availability of value addition technologies for proposed crops. Access to market linkage for value added products. Access to primary processing facilities. Access to the storage facility. Awareness about quality standards of value added products of proposed crops.

Agro-ecological Assessment : Sustainable use of land resource needs a set of agro-ecological conditions for their optimal growth. Optimum use of land requires that land resources be well characterised and their capacities for all likely uses including agriculture, forestry, grazing and other uses determined.

Bioclimate : The mean annual rainfall over Rajasthan varies from less than 100 mm in the extreme western parts of Jaisalmer district to more than 900 mm in the eastern parts of Jhalawar and Banswara.

The mean annual temperature ranges between 24°C and 27°C. In the western part, average temperature, as recorded in the months of May and June is around 40°C to 43°C, while the mean winter temperature drops to 12.9°C during the months of December and January. Potential evapotranspiration refers to total water losses through evaporation from soil and water bodies and transpiration from plants. The annual potential evapotranspiration values of Rajasthan computed according to Pennman's formula vary from less than 1300 mm. In the districts of Dungarpur and Banswara in eastern Rajasthan to more than 2000 mm in Jaisalmer district in the western Rajasthan.

Agro-ecological Assessment-Suitability for Crops, Trees, Grasses

Land suitability for different crop is assessed by matching the land characteristics of mapping unit with crop requirement at different limitation levels. Limitations are deviations from optimum condition of land characteristics, which may adversely affect the kind of land-use. If land characteristics are optimal for plant growth, then it had no limitation. On the other hand, when the same characteristics is unfavourable for plant growth, then it has a severe limitation, thus the evaluation is being done by comparing the land characteristics with the limitation level of requirement.

Agro-ecological Unit, Udaipur district

AEU map of Udaipur district has been prepared by super imposing the variability of rainfall (10 cm) and LGP (15 days) on soil map on 1:250,000 scale and 28 soil mapping units have been delineated as 72 AEU. They are described in terms of soil site characteristics, rainfall, length of growing period, available water capacity and assessed for their suitability for different crops.

Landuse Option

The potentially suitable area for sorghum, oilseed crops (castor, sunflower) and pulses is higher as compared to maize. Introduction of these crops can increase cropped area substantially from the present 16.7%. Suitability of wheat, barley and mustard is almost same under iriigated condition in normal soil, whereas barley and mustard has higher suitability in saline soils. Gram has relatively low suitability in soils of ER upland plain in Mavli and Vallabhnagar tehsil and part of Kherwara tehsil. Soils of Dhariawad, Gogunda, Sarada, Salumbar have moderate suitability. Valley soils are moderately suitable under irrigated condition for wheat, barley and mustard. There is not much scope for increasing the area in rabi. However, there can be shift from wheat to barley, mustard, gram and rabi pulses. Associated soils with rock outcrops in Salumbar, Sarada, Dhariawad and Girwa tehsil are moderately suitable for irrigated crops. Soils of Kherwara, Salumbar, Sarada, Gogunda and Girwa tehsil are moderately suitable, whereas soils of Kotra and Jhadol tehsil are of low suitability. In valley soils in Jhadol and Gogunda tehsil, there is possibility of augmenting water resources through watershed development and creating water storage structures. Development of pastures and plantation in pockets in shallow skeletal soils and rock outcrops can substantially enhance the recharge of groundwater.

Utilization of harvested water

There are 10 decision sets showing the combination of different crops, which is proposed in watershed area for utilization of harvested water. The each crop has been allocated area on the basis of minimum number of irrigation required. Among the ten combinations, the one which shows the maximum benefit is considered appropriate for allocating area to the different crops wheat, gram and mustard.

Future Line of Work

Validation of LUP at farm Level : To ensure landuse plans are sustainable and reliable at the level of end users, the same need to be validated at farm level when land resource programme at village level is operationalized involving the state departments, state agricultural universities and sister organizations.

A dynamic DSS will be developed to address issues related to needs of the stakeholders, seasonal climatic aberrations and market needs.

To develop interactive Multiple Goal Linear Program (MGLP) incorporating employment generation and water use efficiency as objective function along with existing goals for area allocation to different components of agriculture to enable policy-makers to analyze landuse options and their consequences.

❑❑❑

Chapter 21

Selection of Suitable Sites for Rain Water Harvesting Structures and Runoff Potential Areas using Remote Sensing and GIS Tools

M.S. Pendke

Agricultural Engineer, AICRP for Dryland Agriculture, Parbhani, (M.S.)

Introduction

Water is also a prime natural resource and considered as a precious national asset. Land and water are the life sustaining resources of the world which are undergoing a rattling stress as a result of increasing anthropogenic pressure on these finite resources for the primary needs and thereby affecting the future food security. Rainwater harvesting in water harvesting structures for supplemental irrigation is an important strategy for stabilizing yields of rainfed crops. Surface water harvesting structures are promoted in the Integrated Watershed Management Programme (IWMP) and under MGNREGS as a drought proofing strategy. More recently, farm ponds are integrated with micro irrigation programs under National Mission, particularly in states like Maharashtra and Andhra Pradesh.

Geographical Information System (GIS) has multiple meanings and the widely accepted views are the effectiveness of map processing, databases and spatial analysis with the help of human expertise. GIS integrates geographically referenced spatial and non-spatial data required at different scales and times, and in different formats. GIS provides a digital representation of the watershed/basin characterization used in the hydrologic modeling. During the systematic survey of watershed/basin, factors like physiography, soils, vegetation, landuse, slope, drainage pattern, etc. are considered simultaneously and huge amount of attribute data are required to be collected. With the advent of advanced computerized technology, the above said problems are best handled through GIS techniques. One of the major advantages of GIS is its capability to overlay multi-thematic data, which could be used in hydrological models, integrated watershed management planning and groundwater assessment study.

Methodology

Data Collection : The SOI toposheet (1:50000 scale) and LANDSAT image (1977).

Table 1 : Satellite data required

Satellite , Sensor	Path/Row	Date of acquisition
Resourcesat-2, LISS IV	2530/060	15th October, 2011
Resourcesat-2, LISS IV	5258/060	24th April, 2012
Cartosat-1	517/311	3rd March, 2011
Resource Sat-2, LISS-IV	98-59-A	25th March, 2016
Resource Sat-2, LISS-IV	98-59-A	20th November, 2016

- The basic soil depth, soil texture and hydrological soil group maps on 1: 250000 scales.
- The rainfall data of last 30 years
- The information on village wise cropping pattern and area

Observations to be recorded

1. Pre and post-monsoon water levels in open wells and borewells.
2. Drawdown in open wells during pumping test.

Drainage and Water Bodies Mapping

Drainage delineation can be done using manual methods using existing maps and high-resolution satellite imageries or semi-automated methods using a Digital Elevation Model (DEM). In this method, reference drainage is extracted from existing topomaps on 1:50K, and the same is updated by overlaying on the latest high-resolution geo-referenced satellite imagery. High resolution satellite imagery gives latest information on the present course of the streams as well as some of the very minor 1st order streams can also be delineated, which are not marked on the existing maps. Water bodies can be easily mapped directly using the latest satellite data.

Methodological Flow Chart

The methodology that should be adopted for Selection of Suitable Sites for Rain Water Harvesting Structures & Runoff Potential Areas using Remote Sensing & GIS Tools is as below:

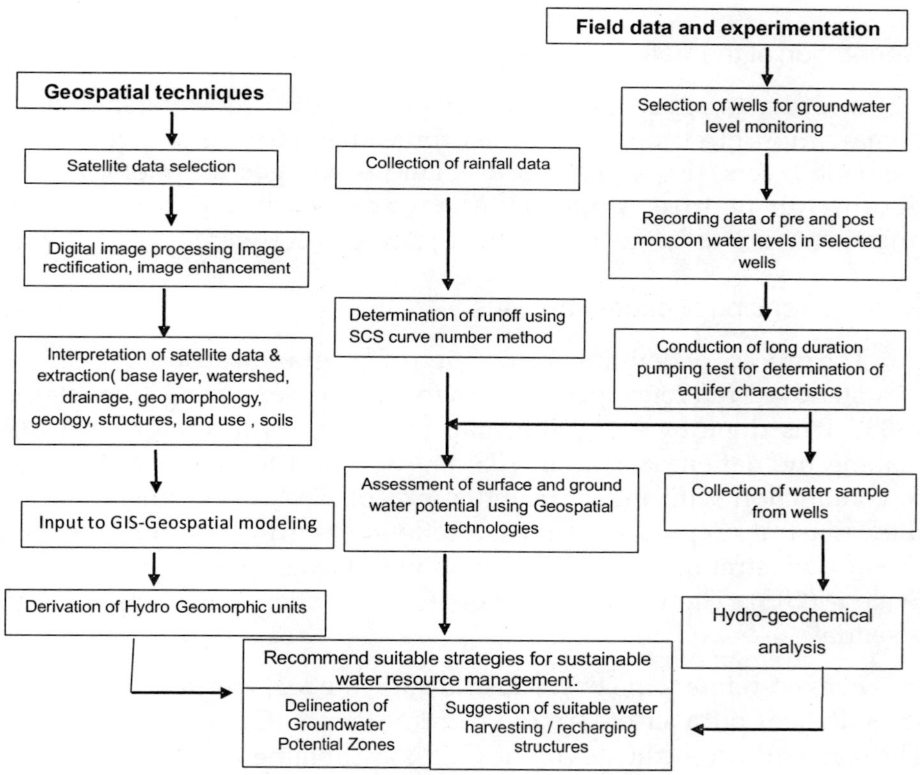

Fig. 1 : Flow-Chart of research methodology

Watershed and Sub-Watersheds Delineation

Watershed delineation can be done by manual method or semi-automatic methods using DEM as input. In manual method, the drainage lines and contours are taken as the reference. First a point is chosen as the watershed outlet, which will be the pore point for all the water draining out of that watershed. Watershed boundary is delineated by drawing lines perpendicular to the elevation contours for the land that drains to the point of interest. The methodology for semi-automatic delineation of watersheds involves the generation or extraction of the DEM for the study area, elimination of the sinks and peaks to generate a depression less DEM, derivation of Flowdirection Grid and the Flowaccumulation Grid, defining the pour point and identifying the ridge lines based on the flow accumulation and automatic delineation of the watershed boundary (Sreenivasan *et al.*, 2010, Sreenivasan and Krishna Murthy, 2017).

Generation of thematic layers

In order to assess groundwater potential in the watershed and demarcation of suitable locations for water harvesting structures, thematic layers/maps like landuse, land cover, geomorphology, soil depth, soil texture, slope, DEM, hydrological soil groups and lineaments are to be generated using Arc GIS software.

Geo-referencing of cadastral maps

The process of defining the position of geographical objects relative to a standard reference grid is known as geo-referencing (Bernstein, 1983). It is done by assigning map co-ordinates to image data for making the data amenable to GIS analysis, and to facilitate viewing in conjunction with maps. In India, except parts of Northern India, the cadastral maps do not have geodetic co-ordinates, limiting the use of Cadastral maps in conjunction with Geographical Information System (GIS). Therefore, geo-referencing of cadastral maps is very essential.

For geo-referencing the cadastral map, the real world co-ordinates for sufficient number of Ground Control Points (GCPs) are required. The real world co-ordinates of the GCP's are obtained through primary sources or secondary sources. The primary source consists of three modes, viz., ground control survey, topographical maps and co-ordinates obtained from GPS.

The merged product of Carto SAT-1 and LISS-IV ortho-images should be used for geo-referencing of the cadastral maps. The image depicted field bunds, cart tracks, small nullahs, settlements, tanks and other cultural features like roads, railway network and canals. These features facilitated identification of ground control points for developing transformation model for image-to-image tie down. The vector format of the cadastral maps already available were taken for geo-referencing using the high-resolution satellite data mentioned above. The geo-referencing of digital cadastral maps consists for first acquisition of GCP's, and then using these GCP's a transformation model was developed, suing which the vector cadastral maps are geo-referenced.

Assessment of Surface Runoff Potential

The rainfall data for the last 30 years should be collected and the rainfall data of the normal rainfall year should be used to estimate the surface runoff potential using SCS curve number technique. The SCS curve number techniques is based on recharge capacity of the watershed. The recharge capacity is determined by antecedent moisture condition and by physical characteristics of the watershed. Antecedent moisture condition (AMC) is used as an index of watershed wetness.

Well Inventory

The open and borewells at spread locations should be selected for monitoring the water level fluctuations. The positions of wells should be monitored by GPS to note the latitude and longitude.

Aquifer parameterization

Evaluation of aquifer characteristic is the prime task for harnessing an aquifer for optimum results yielding information about how much groundwater is available for development and what will be the consequences of withdrawing a certain quantity of ground water. Aquifer parameters i.e. transmissivity (T) and specific yield (Sy) are often determined by means of pumping test. Scientifically planned pumping tests provided information for solution of many regional, as well as local, groundwater flow problems. It also provides information about the yield and drawdown of the well, which in turn is essential for selecting the type of pump, estimating cost of pumping, well efficiency, etc.

Water level fluctuation method

The water level fluctuation method is based on the application of groundwater balance equation, which is stated in general terms as follow for any specified period.

Input – Output = Storage increase

In the above equation, the terms input and output are used in general sense, referring to all components of groundwater balance, which are either input to the unit, or output from the unit of groundwater system taken up for resource assessment e.g. watershed, block. Hence, input refers to the recharge from rainfall and other sources and subsurface inflow into the unit. The output refers to the groundwater draft, groundwater evapotransipiration, base flow to streams and sub-surface outflow from the unit. The above equation hold good for any period and hence, it can be applied to the year as a whole or to different seasons in the year separately. Watershed with well-defined hydrological boundaries is an appropriate hydrological unit for groundwater resource estimation. In hard rock areas, the hydrogeological and hydrological units normally coincide, thus this method is very appropriate. (Ministry of Water Resources, GOI, 2009).

Computation of groundwater recharge

Groundwater recharge is the process by which water percolates down the soil and reaches the water table, either by natural or artificial methods. Quantification of the rate of natural groundwater recharge is a pre-requisite for efficient groundwater resource management. It is particularly important in the region with large demand for groundwater supplies, where such resources are the key to economic development. The amount of water that may be extracted from an aquifer without causing depletion is primarily dependent upon the groundwater recharge. However, the rate of aquifer recharge is one of the most difficult factors to measure in the evaluation of groundwater resources.

This method linked the change in groundwater storage with resulting water table fluctuation and then recharge is equal to change in groundwater storage.

$$\Delta S = S_y \times \text{“} h$$

Where,

ΔS = Groundwater recharge, cm

Δh = Difference in water level, cm, and

S_y = Specific yield

The groundwater recharge for the watershed was determined and then by interacting the village boundary map, sub-watershed map and groundwater recharge map, the groundwater recharge for each sub-watershed was determined and map was prepared using GIS technique (Karanth, 1987, Todd, 1980)

Delineation of Groundwater Potential Zones

Groundwater potential zones mapping was achieved through the use of remote sensing techniques and GIS analysis. The mapping of groundwater potential was based on the basic principles of occurrence and flow of groundwater in the sub-surface. The controlling factors for the groundwater occurrence and movement were the nature of the sub-surface litho-units, the amount and spatial variation of the weathering and the structural fabric mainly dictated by the joints/ fractures (lineaments). The remote sensing data with sufficient ground truth information were used for deriving information on the geology, geological structures, geomorphology, recharge conditions (defined by the nature of the subsurface material and hydrological conditions), which ultimately defined the groundwater regime of an area. The multi-criteria decision-making technique is GIS was further used for arriving at the groundwater potential of the watershed. The thematic information derived from remote sensing techniques was integrated in GIS and suitably analyzed along with field well inventory data, such as groundwater levels, well yields, etc. The above mentioned thematic layers were integrated using UNION tools in ArcGIS. GIS analysis provided the output groundwater prospect maps, showing the probable zones/regions with different groundwater prospects (Sreenivasan *et al.*, 2010). Based on the well yield data of pumping test wells, the groundwater potential zones should be demarcated on satellite image and groundwater potential zones should be classified as poor, moderate, good, very good and excellent.

Water Balance Studies and Water Resources Budgeting

An equation that describes the annual water budget of a natural system like watershed states that water input (I) equals discharge (D) plus or minus changes in water in storage (ΔS):

$$I \text{ (water entering system)} = D \text{ (water leaving system)} + \Delta S.$$

In natural systems, water enters the system as precipitation (P) and leaves the system as stream flow (SF) (surface runoff plus ground-water discharge to streams) and evapotranspiration (ET)

P = SF DS + ET

where

P = precipitation,

SF = stream flow leaving the watershed,

DS = change in storage, and

ET = evapotranspiration.

Total available water

Available water = Area. in ha x Rainfall in m

Water stored in the soil 5%

Water percolated in soil10%

Soil moisture 30%

Total required water

Population x required water 50 lpd x 365 days

Water required for other purpose (2% of rainfall)

Total available water-Required water = Water surplus/deficit

Considering the crop water requirement of each crop, the total present water requirement of the kharif crop, rabi crop including plantation in each village under each sub-watershed should be worked out. Similarly, the input from the rainfall should also worked out. The various losses, such as runoff, evaporation from water bodies should be estimated and accordingly the water available in the watershed should be estimated. Further, the water availability should be compared with total water demand (crop and population) and thus the water deficit or surplus in each watershed should be estimated. Based on these data source, the geospatial maps can be generated.

Water Resource Management

The goals of water resource management project may be achieved by planning suitable water harvesting structure at proper locations. Water resource planning techniques are used to determine what measures should be employed to meet water needs and to take advantage of opportunities for water resource development and also to preserve and enhance natural water resources and related land resources. Water resource projects are constructed to develop or manage the available water resources for different purposes. According to the National Water Policy (2002), the water allocation priorities for planning and operation of water resource system should be for the domestic needs, irrigation and industries. If the planning is on watershed basis, then generally the water available within the basin is to be used unless there is a inter basin transfer.

Water resource development action plan

The water resources development action plan mainly suggests locale specific water conservation and development measures to be taken up within the watershed. The development measures include the suggestions on the type of groundwater abstraction structures suitable in an area based on the nature of the sub-surface formation, the landforms and the groundwater dynamics in an area (Sreenivasan *et al.* 2010, Sreenivasan and Krishna Murthy, 2017).

The water conservation measures include the area specific suggestions about the type of water harvesting and groundwater recharge structures to be taken up under the limited constraints of the various natural resources criteria and the socio-economic feasibility. Water harvesting is carried out mainly to store rain water that goes as the runoff. The main objectives of surface water harvesting are: storing excessive run off for future use; augment percolation/recharge to aquifers; checking of floods and checking of land erosion. Water harvesting measures must be implemented on a watershed basis for effectiveness and for better management (Sreenivasan *et al.* 2010, Sreenivasan and Krishna Murthy, 2017).

Appropriate location of water harvesting structures is important for effectiveness of these structures. There are number of parameters which need to be considered for selecting sites for a water harvesting structure. The essential parameters which control the location and type of the water harvesting structures are: nature of subsurface soil and rock strata and their thickness, extent of weathering and lateral variations, geological structures, depth to bed rock, topography/slope, geomorphology, drainage characteristics and landuse pattern in an area. The information on many of the above mentioned parameters can be easily derived from the remote sensing satellite data, as discussed in the earlier paragraph (Sreenivasan *et al.*, 2010, Sreenivasan and Krishna Murthy, 2017).

Selection of suitable sites also depends on the site conditions like subsurface geology and soils, permeability of the subsurface material, the stream/drainage conditions, catchment area of the structure, etc. Depending on the type of water harvesting method and terrain conditions number of water harvesting structures are prevalent, such as farm ponds, minor irrigation tanks, check dams, nala bunds, percolation tanks, recharge pits, underground bhandaras, recharge wells, etc. (Sreenivasan *et al.* 2010, Sreenivasan and Krishna Murthy, 2017).

Based on the water deficit in each sub-watershed and by integrating the thematic maps of geomorphology, landuse, soil texture and land slope by UNION tool in arc GIS, the suitable location of water harvesting structures were decided preferably on the 1st and 2nd order streams and marked on the satellite image along with its latitude and longitude. Based on the present water deficit in each subwatershed and considering the geomorphology and landuse, the water harvesting structures like cement plugs and percolation tanks are suggested for improving the surface and groundwater availability. In addition to this, number of farm pond are suggested to be undertaken by the farmers on their respective field to meetout the seasonal water requirement of the crops through supplemental/ protective irrigation to crops.

Identification of rainwater harvesting and groundwater recharge sites

The methodology for locating sites for water harvesting involves a set of steps involving interpretation of information from remote sensing data for generation of all the necessary thematic layers and further encoding them on to a GIS platform, and then modelling in GIS as per criteria requirement for specific water harvesting structure (Sreenivasan *et al.*, 2010). Most of the water harvesting structures was preferred along streams, therefore the first step in GIS will be the extraction of buffer area along the second and third order streams. The selection of the order of the stream again depends on the type of structure. All the thematic layers were then clipped for the buffer extent. Drainage of lower order was not selected, as these drains will not be able to give required volume of water. A criteria was developed for locating various water harvesting structures. The next step was to assign weights for different feature classes in the respective thematic layers based on the criteria developed for different structures. Different weights can be given for water harvesting and recharge sites. After assigning weights to all the layers, the layers were integrated to generate the Composite Mapping Units (CMU). The weights within each CMU (indicating the different bio-geophysical parameters for that homogeneous unit) were added upto arrive at the cumulative weighted index. Prioritization of the sites was done by grading the cumulative weights into priority classes based on the need. Final selection of sites was done by overlaying the drainage layer on the final cumulative weighted layer. The structures should be located at places where the sites qualify the criteria with highest grades" (Sreenivasan *et al.*, 2010, Sreenivasan and Krishna Murthy, 2017).

Suggestion for suitable location for soil conservation structures

The life and storage capacity of the water harvesting structures depends on the treatment of the upper catchment of the watershed area. Thus, to avoid the further siltation in proposed water harvesting structures, a soil conservation plan should be suggested and accordingly the locations for nala plugs, loose boulder structures, gabion structures, etc. are to be suggested on 1st and 2nd order streams in the watershed.

References

1. Eastman, R. J. 2001. IDIRISI Andes; Guide to GIS and Image Processing., Clark University, USA., pp. 144.
2. Sahu, B.K. 2001. Groundwater Modelling in hard-rock terrain. In: Proceedings of national seminar on groundwater resources, MLS University, Udipur, India, pp. 16-21.
3. Saraf, A.K. and P.R. Choudhury. 1998. Integrated remote sensing and GIS for groundwater exploration and identification of artificial recharge sites. International Journal of Remote sensing, 19(10): 1825-184.

❑❑❑

Chapter 22

Role of Remote Sensing and GIS Techniques for Groundwater Recharge Site Selection and Runoff Estmation in Hard Rock Areas – A case Study from South India

M.S. Shirahatti

Chief Scientist, Regional Agricultural Research Station, Vijayapura

Introduction

Groundwater is a dynamic and replenish able natural resource but in hard rock terrains availability of groundwater is of limited extent. Occurrence of groundwater in such rocks is essentially confined to fractured and weathered horizons. In India, 65 per cent of the total geographical area is covered by hard rock formations with low porosity (less than 5 per cent) and very low (10^{-1} to 10^{-5} m/day) permeability (Saraf and Choudhary, 1998 and Sahu, 2001). Therefore, efficient management and planning of groundwater in these areas is of the utmost importance. Proper scientific planning and management require immense data to make predictions of water availability.

Materials and Methods

Study area

The present study was taken up in the upper Don river basin, Bijapur district of Karnataka state (Fig. 1). Its catchment is bounded by $10^0 42'8''$ to $16^0 59'26''$ N and $75^0 12'24''$ to $75^0 39'40''$ E. The major portion of the catchment falls in Bijapur taluka only. The physiography is gently sloping and there is the total relief of 143 m, normal rainfall of Bijapur is 590 mm and minimum and maximum temperatures recorded were 9.48^{0}C and 41.75^{0}C respectively. While the monthly reference evapotranspiration (ETo) was in the range of 82.77 to 258.23 mm. The major part of the upper Don watershed is occupied by basalt and other trap intrusions. Black cotton soil is dominant type of soil and it is followed by mixed loamy soil.

Thematic map generation

In this study, nine different thematic layers were used. Among the nine layers, six layers viz., geology, soil, landuse/cover, water table fluctuation, depth to bedrock and slope thematic maps were generated from the different data sources. For developing slope map, SRTM data was used. Other three thematic maps, viz., drainage density, lineament density and geo-morphology were derived from the above maps used. For the development of landuse/cover map the IRS ID LISS-III digital data was used and by digital image processing, technique, totally eight different landuse/land cover classes were demarcated in the study area. The high transformed divergence (TD) values among the classes were observed. It shows that these classes were spectrally separable. After developing different thematic maps, under the GIS environment geo data base was developed.

Modelling for site selection

To meet a specific objective, it is frequently the case that several criteria will need to be evaluated. Such a procedure is called Multi-Criteria Evaluation (Carver, 1991). Another term that is sometimes encountered for this is modelling. The artificial recharge process is presented in the form of flow diagram in Fig.2. density and geo morphology were derived from the above maps used. For the development of landuse/cover map the IRS ID LISS-III digital data was used and by digital image processing, technique, totally eight different landuse/land cover classes were demarcated in the study area. The high transformed divergence (TD) values among the classes were observed. It shows that these classes were spectrally separable. After developing different thematic maps, under the GIS environment geo data base was developed.

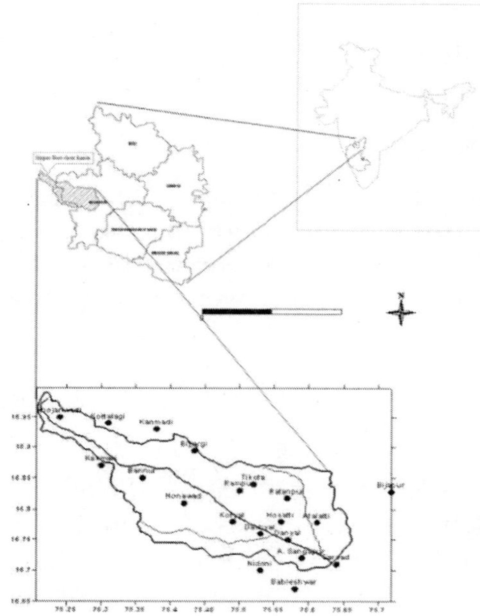

Figure 1: Location map of the study area

Modelling for site selection

To meet a specific objective, it is frequently the case that several criteria will need to be evaluated. Such a procedure is called Multi-Criteria Evaluation (Carver, 1991). Another term that is sometimes encountered for this is modelling.

Weighted linear combination (WLC) model or (Indexing overlay)

Comparing weighted linear combination (WLC) models with Boolean Models executive routines, it is identified that WLC has more flexibility and ability for priority indication on spatial units of factor maps (Bonham Carter and G.F., 1991). In this type of models, it is possible to consider knowledge driven weighting. With respect to mentioned characteristics, this model is relatively simple and more realistic, hence this model was used for artificial recharge site selection process. With weighted linear combination (WLC), factors are combined by applying a weight to each followed by a summation of the results to yield of suitability map (Eastman, 2001) i.e.;

$$S = \sum wi\ xi \text{ ----------- } (1)$$

where,

wi = The weight of ith factor map

xi = Criteria score of class of factor i

S = Suitability index

In the present study, all the thematic layers were integrated in IDIRISI GIS platform in order to prepare a map depicting suitable areas for artificial groundwater recharge. The total weights of each pixel of the final integrated layer were derived from the following equation:

$$S=(GM_w\ GM_{wf} + G_w\ G_{wf} + S_w\ S_{wf} + L_w\ L_{wf} + SL_w\ SL_{wf} + LD_w\ LD_{wf} + WF_w\ W_{wf}F + DB_w + DB_{wf}) \text{----------}(2)$$

Where, 'w' represents the weight of each factor, while 'wf' represents the weight of the each class of the individual factor, namely geomorphology (GM), geology (G), soil (S), landuse/cover (L), slope (SL), lineament density (LS), water table fluctuation (WF), depth to bed rock (DB). Thus, the artificial groundwater recharge index 'S', which is dimensionless quantity that helps in indexing the probable groundwater recharge zones in the area, was estimated using equation (2) for each pixel in the final integration layer and was re-grouped into different classes with equal class interval to divide the entire study area into different artificial recharge zones. In the similar studies at West Bengal (Choudary *et al.*, 2006) used the same procedure.

Model evaluation

Accuracy of models is another parameter that is important for selecting optimum model. Therefore, we need a criterion for comparing the accuracy of output model results. There are several criterion in groundwater recharge site selection for validation of the models, some of them are well yield data , geo electrical resistively data and the well log analysis. However, in this method the validity of the model developed was checked against the bore well yield data, which reflects the actual groundwater potential. Then, by comparing the location status value and output maps estimated status, each correct estimating was identified by " +" and each incorrect estimating was identified by "-".

Results and Discussion

Geology

The study area comprises mainly of basalt and other trap intrusions. Detailed analysis revealed that there are six types of geological classes found viz., laterites, sparse to moderate prophyrtic, sparse to moderate porphyrotic (compressed), high to moderate prophyrtic, mega cryst and moderate to high prophyrtic. The geological formations like laterite is considered better than the other geological formations for the recharge occurrence.

Soils: The major part of the study area is covered by vertisols (black cotton soil) and remaining part is with loamy soils (Fig.3). There are, however, variations in the type of soil and different varieties of soil viz., very fine montmorillonitic (chromosters), fine montmorillonitic, loamy mixed lithic and very fine montmorillonitic. Recharge point of view, loamy mixed lithic soil is considered better than the montmorillonitic soil group.

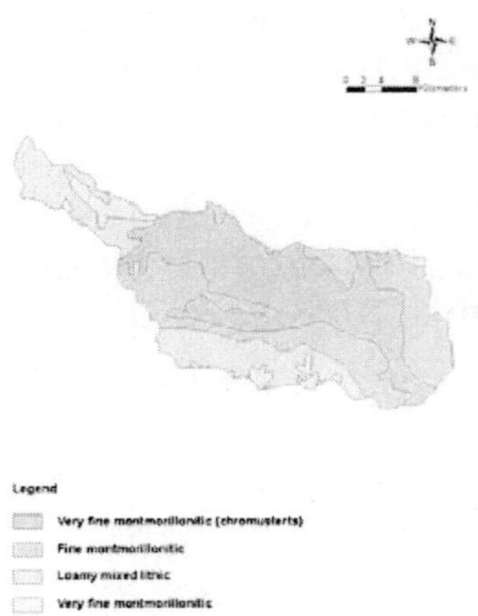

Figure 3: Soil map of the study area

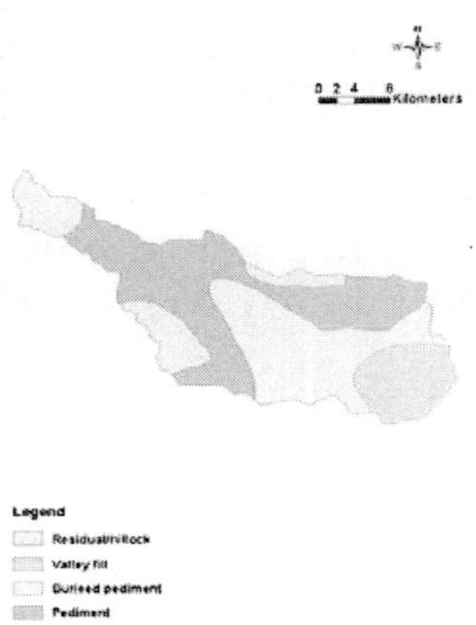

Figure 4: Geo- morphology map of the study area

Geo morphology

The region has gently sloping topography. The major geomorphic landforms demarcated are pedimented plains, buried pediments, valley fills and residual/hillocks (Fig. 4). Further, on recharge point of view, valley fill, buried pediment and pediment plains are favourable, while residual hills are least favourable.

Landuse / Land cover

There are totally eight different land use/cover types were demarcated and shown in Fig. 5. The separability analysis and accuracy assessment was also carried out. The overall assessment of about 85 per cent was observed. In view of the recharge, cultivable lands, orchard and summer crops were given higher weightage, whereas, settlement, wastelands were given low weightage.

Slope

The major part of the area is having gentle slope. However, variations in the slope (0 to 7.31 per cent) was made to group the entire area into four slope classes i.e., 0-1,1-3, 3-5 and > 5 per cent. The areas having slope 0-1 and 1-3 per cent were assigned higher values and the areas, which are having slope more than 3 per cent were considered as the moderate in view of the ground water recharge.

Lineament density map

In the study area, the lineament density was in the range of 0 to 2.96 Km/ sq.Km. Hence, for the analysis they were grouped into four classes viz., > 1.5 (high), 1 to 1.5 (moderate), 0.5 to 1 (low) and < 0.5 (very low) Km/ sq.Km respectively (Fig. 6). Further, for recharge point of view, more weightage was given where lineament density was higher than 1.5 Km/ sq.Km, whereas, low weightage was given where the lineament density was lower than 0.5 Km/ sq.Km.

Land use classes

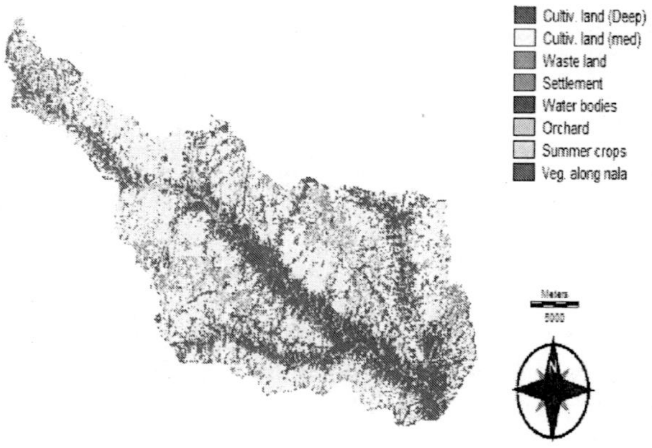

Figure 5: Land use map of the study area

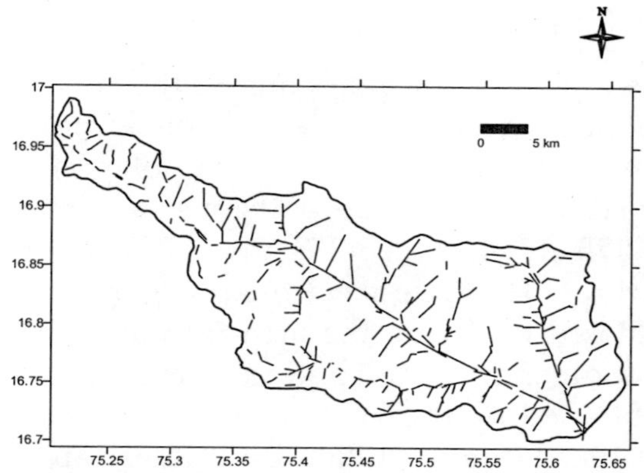

Figure 6: Lineament map of the study area

Drainage density map

The dominant drainage pattern observed is dendritic to sub-dendritic. A glance at drainage density map reveals that density values ranged from 0 to 6.32 Km/Sq.Km. For the analysis purpose, they were re-grouped into four categories, i.e., high (> 3.5), medium (2-3), low (1-2) and very low (0-1) Km/Sq.Km. Considering recharge point of view, more weightage was assigned to low drainage density regions, whereas low weightage assigned to high drainage density regions.

Watertable fluctuation

The fluctuation was in the range of 0 to 13.50 m. Except, small patches (< 3.78 per cent of the total area) on the lower and northern side of the watershed, the fluctuation was in the range of 0 to 6 m only, while in the patches mentioned above, it was upto 13.5 m. For selecting artificial recharge sites, the maximum weightage was assigned to the regions with water table fluctuation more than 4 m, whereas less weightage was given where fluctuation was less than 4 m.

Depth to bedrock map

The depth to bedrock was in the range of 15 to 77.78 m. For the analysis purpose, the depth to bedrock was re-grouped into four classes, viz., < 15 m, 15–30 m, 30–45 m and > 45 m. For assigning weightage for groundwater recharge site selection, the regions with depth to bedrock more than 30 m was given high weightage, whereas, for the less than 30 m depth given the low weightage.

Data integration for suitability analysis

The final weights of the polygons in the final integrated layer were obtained using equation 2, which is considered as the groundwater recharge index of individual polygons. The total range of the weights (150 to 300) of the polygons were classified into four classes, viz., < 175, 175-225, 225-275 and > 275 by dividing the total class range by three. The weights of the polygons coming under each range were clubbed to divide the area into four different classes/ zones viz., poor, moderate, good and excellent. The artificial recharge map, thusm obtained by the WLC method is shown in Fig. 7. The areal extent of the different suitability classes are fall in the good zone, followed by moderate (48%), poor (0.93%) and excellent (0.57%) zones.

The good zone occupied most of the lower and small patches in the upper parts of the watershed. This area was occupied by buried pediment landform with low to moderate slope (0-1 and 1-3%), moderate WT fluctation (4-6m) and moderate to high aquifer thickness. The moderate zone was occupied by most of the upper and small patches in the lower part of the watershed. This area was covered by pediment and parts of residual hill type land forms, moderate to high slope areas and WT fluctuation was also relatively low. Further, the poor zone was found on the south western side and upper part of the watershed. This area was occupied by residual hillock land forms, slope is relatively high (> 5%) and WT fluctuation was low (2 to 4m). Further, the sub-watershed wise various suitability classes are presented in the Table 1. A close look at data reveals that the most of the sub watersheds were dominated by either good or moderate suitability zones.

Evaluation of models

The validity of the models developed was checked against the borewell yield data, which reflects the actual groundwater recharge potential. Yield of borewell ranged from 68.4 to 211.2 lpm. For the evaluation, the yield is in the range of > 200 lpm was put under excellent zone, 140 to 200 lpm in good zone, 80 to 140 in moderate zone and less than 80 lpm under poor zone. Then, by comparing location status value (well yield ranges) and output maps estimated status, each correct estimate was identified by "+" and each incorrect estimate was identified by "_"This is the fairly good estimation, therefore, it is proposed that for the study area, WLC model is an optimum model for artificial recharge site selection.

Runoff Estimation

For assessing the water resources of the study area, Thiessen polygon was drawn from five rain gauge stations (Tikota, Bijapur, Jatt, Bableshwar and Telsang) and for the analysis, ten years (1998-2007) daily rainfall data was used. For the micro-level planning, the entire study watershed (643.5 Sq.Km) was sub-divided into ten sub-watersheds. The Soil Conservation Service-Curve Number method was developed from many years of stream flow records for agricultural watersheds in several parts of the United States. For the Indian conditions, with certain modifications, Soil Conservation Department has proposed following equations for different region of India:

$$Q = \frac{(P-0.3S)^2}{P+0.7S} \text{ --------(1)-(For other soils)}$$

$$Q = \frac{(P-0.1S)^2}{P+0.9S} \text{ --------(2) -(For black cotton soil)}$$

Further, The curve number (CN) is calculated by ;

$$CN = \frac{25400}{S} - 254\text{-----(3)}$$

where, P is precipitation, S is recharge capacity and Q is the direct runoff in depth and all are in same units.

i) Landuse/cover map : For the development of landuse classes map, the IRS ID LISS-III digital data was used as per the requirement of curve number method .

ii) Hydrologic soil group map : Hydraulic soil group map was generated by assigning appropriate hydraulic soil groups to the various soil classes in the soil map prepared by the National Bureau of Soil Survey & Land Use Planning (NBSS & LUP).

iii) Curve number map : Curve number map was generated using both landuse and hydraulic soil group map using CROSS TAB module in the IDIRISI GIS environment.

iv) Thiessen polygon map : Rainfall distribution map is one of the important inputs for the runoff calculation. The THIESSEN module of the IDIRIS software was used to generate the Theissen polygon using the five rain gauge stations, situated in the study area.

Loose coupling model for runoff computation:

In loose coupling model, the GIS and simulation model represent two distinct software packages and the data transfer from one to another is made by storage of data in one system and subsequent reading of the data by the other. The GIS software is used to pre process and post process the spatial data. Accordingly, here loose coupling (models linked to GIS) was used, after developing the curve number in the GIS environment, all the basic data which is required for runoff calculation viz., curve number, soil unit and rain fall zones were taken out from the GIS environment and used the same for the input for the runoff calculation programme, which was coded in the BASIC language. The output (runoff values) from the programme was once again transformed pixel by pixel to the GIS environment, and further analysis and display was carried out in the GIS.

Results and Discussion

The Runoff data analysis revealed that maximum runoff occurred during the June, September and October months. Further, the runoff as the percentage of annual rainfall varied from 5.08 to 25.66 per cent with average of 14.38 per cent (81.6mm) of annual rainfall. The runoff yield would be moderate to good at the rainfall of probability level of 50 and less. Further, the yearly average of sub-watershed wise annual runoff yield was in the range of 121.33 to 927.44 ha m, with an average of 488.88 ha m. The monthly the rainfall–runoff relationships were established for predicting the runoff on the basis of monthly rainfall. It revealed that a certain minimum rainfall (19.1 to 43.6 mm) was required for initiating the runoff.

Conclusion

Satellite remote sensing IRS ID LISS-III digital data can be used successfully to classify the different landuse/land covers, and Shuttle Radar Topographic Mission (SRTM) digital elevation model (DEM) data can be used to draw the classified slope maps. Multi Criteria Evaluation

(MCE) based integrated models such as weighted linear combination (WLC) can be used successfully to locate suitable sites for the artificial recharge.

References

1. Bonham Carter and G.F., 1991. Geographic Information System for Geoscientists: Modelling with GIS. Pergamon, Ontario, pp. 319-470.
2. Carver, S.J. 1991. Integrating multicriteria evaluation with Geographical Information Systems. International Journal of Geographical Information Systems, **5**: 321-339.
3. Chowdhury, A., Madan, K. Jha, V.M. Chowdhary, and B.C. Mal. 2006. Selection of artificial recharge zones in West Midnapur district of West Bengal, India: A GIS approach. **In**: An International Prospective on Environmental & Water Resources, Organized by Environmental and Water Resources Institute of ASCE, New Dehli, pp. 1-8.

❑❑❑

Chapter 23

Dynamic Crop Planning for Proactive Monsoon Management under Rainfed Regions

A.K. Kothari, J.K. Balyan and Devyani Dodia

Dryland Farming Research Station, Post Box No. 62, Arjia, Bhilwara, (Maharana Pratap University of Agriculture and Technology, Udaipur)

Introduction

Rainfed agriculture occupies 68% of country's cultivated area and supports 40% of human and 60% livestock population. The region contributes 91% of coarse cereals, 91% of pulses,80% of oilseed and 65% of cotton production in India. Most of the studies are undertaken at district while the data on rainfall, crop and soil are available at micro level. Swaminathan (2005) suggested to develop contingency cropping plans and climatic managers at micro (village) level for proactive monsoon management. "Meteorological approach" is used in most of the studies for crop planning. While, it has been suggested to identify and quantify seasonal rainfall variability and its related risks with variability of onset of monsoon for managing the climatic risks in crop production (Stewart, 1991).

Site Selection

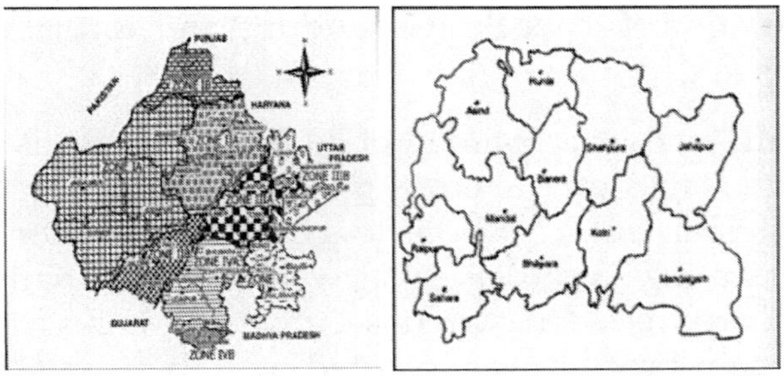

Figure 1: Location of study area of Rajasthan

Methodology

The study was conducted in Bhilwara district which lies at 25° N latitude, 75° longitude and altitude of 463.2 m above mean sea level. Daily rainfall data for a period of 57 years (1960-2016) were collected from the Office of the District Collector, Bhilwara in Rajasthan state. The weekly meteorological data recorded at Dryland Farming Research Station, Bhilwara (Research Station) for a period of 32 years (1985-2016) were used for determination of weekly reference evapotranspiration. The region under study receives an annual rainfall of 608.4 mm and ranges from 277.2 mm to 1091.7mm with a coefficient of variation of 31.24 per cent. Almost 90 per cent of annual rainfall is received during rainy season (June to September). The soils of the region are sandy loam to clay loam in texture having shallow to medium depth. The principle crops in the region are maize, groundnut, black gram, green gram, sesame and sorghum fodder during *kharif* season.

Onset and Withdrawal of Monsoon

It was computed by using a modified Morris and Zandstra (1979) method which was validated from the observed onset of monsoon at the Research Station and is described below:

i) The onset of monsoon in a standard meteorological week (SMW) is considered, if the forward accumulation of weekly rainfall is reached to a pre-determined quantity of 100 mm, 75 mm and 30 mm upto 25,26 and onward 26 SMW with a condition that any week having 0.0 mm rainfall than restart accumulation of rainfall from that SMW.

ii) The withdrawal of monsoon in a SMW is considered, if the backward accumulation of weekly rainfall is 10 mm in that SMW from 47th week considering mono cropped area having maximum crop duration of 18 weeks (Groundnut) and normal week of onset of monsoon.

Dry Spell Analysis : Rainfall of 20 mm per week is adequate for all the growth stages of the crops grown in the region. Thus, if in a given week the rainfall received is less than 20 mm, then week can be designated as a dry week else wet week (Pandarinath, 1991). The probability of dry week and two consecutive dry weeks were computed by Markov chain process, as suggested by Robertson (1976).

Moisture Adequacy Index : It was computed as suggested Krishnan (1979) from the following criteria for weekly soil moisture adequacy index:

MAI = AE / ETo x 100.

Classification of MAI on Weekly basis are decided as follows :

i) MAI >= 75 % -Excellent , ii) MAI = 50 to 74 % Good

iii) MAI = 49 to 24% -Poor iv) MAI < 24 % -Very Poor

Moisture status, If MAI = 0 to 49, during active growth stages of the crop, it is considered as drought. Where, AE and ETo are actual and reference evapotranspiration for the period.

Outcomes

The onset of monsoon was determined from the modified Morris and Zandstra method for 57 years (1960-2016) of rainfall data. Rainfall analysis for dry spell and water balance was carried out by 'meteorological' approach and first step of 'Onset of monsoon' approach (considering onset of monsoon as datum). Further, rainfall characteristics, water surplus / deficit and moisture adequacy index (MAI) were also used to determine for different weeks of onset of monsoon from 'Meteorological' and 'Onset of monsoon' approach. The results obtained, thus, are compared for obtaining the more appropriate approach for crop planning.

Characteristics of rainy season

The onset and withdrawal of monsoon is determined from the modified Morris and Zandstra method. Considerable variation in onset and withdrawal of the monsoon was observed from year-to-year. Therefore, the mean, earliest and latest SMW during which the onset and withdrawal of monsoon have commenced are given in Table 1.The data reveal that the normal week of onset and withdrawal of monsoon were observed in 27th and 38th SMW, respectively. While the earliest onset and withdrawal of monsoon was observed in 25th and 33rd SMW, respectively. The onset of monsoon was observed for 1, 11, 17, 17, 9 and 2 year out of 57 years in 25th, 26th, 27th, 28th, 29th and 31st SMW, respectively. The length of rainy season varied from 4 to 19 weeks with a mean of 10.8 weeks.

Table 1: Characteristics of rainy season at Bhilwara

Onset of monsoon, SMW			Withdrawal of monsoon , SMW			Length of rainy season, weeks		
Early	Late	Mean	Early	Late	Mean	Maxi.	Mini.	Mean
25	31	27.5	33	47	38.4	19	4	10.8

Probability of dry week

The initial and conditional probabilities of dry weeks have been determined from 'Meteorological' and 'Onset of monsoon' approaches. The results presented in Table 2 reveal a significant difference in probability levels during initial period of six weeks and thereafter difference is narrowed down. For example, the probability of dry week in 27th SMW (normal week onset of monsoon) is 40 per cent due to 'Meteorological' approach against nil probability due to 'Onset of monsoon' approach. The results obtained are quite different and controversial. The amount of rainfall accumulation for considering the onset of monsoon was more than 30 mm per week, which implies that the results obtained from 'Onset of monsoon' are justified and it also reveals the lacuna of not considering the shift in origin of onset of monsoon in the conventional 'Meteorological' approach. The controversy in the results is due to shift in origin of onset of monsoon only. Similarly, the data presented in Table 2 indicate difference in probability levels for conditional probabilities of dry weeks in different weeks for crop growth period. However, results obtained from 'onset of monsoon' have been considered for crop planning.

Table 2: Initial and conditional probability of dry week from different approaches

Meteorological approach				Onset of monsoon approach			
SMW	P(D)	P(D/W)	P(D/D)	Week	P(D)	P(D/W)	P(D/D)
27	40	26	74	1	0	0	0
28	39	59	41	2	28	100	0
29	30	59	41	3	35	65	35
30	37	57	43	4	40	61	39
31	37	62	38	5	32	56	44
32	28	63	38	6	44	60	40
33	35	55	45	7	42	25	71
34	42	42	58	8	46	46	54
35	54	39	58	9	47	37	63
36	56	41	59	10	58	30	70
37	65	27	73	11	74	33	67

38	86	29	69	12	84	17	81
39	86	10	90	13	89	14	86
40	91	13	87	14	88	8	92
41	93	9	91	15	91	12	88
42	95	7	93	16	93	6	94
43	96	4	96	17	95	4	96
44	100	4	96	18	96	4	96
45	98	0	100	19	98	2	98

Onset of monsoon and water balance parameters

The onset of monsoon is the only known monsoon parameter before putting the seed in the soil and relating it with rainfall characteristics or water balance will be of great importance for agriculturist and farmers. Therefore, mean rainfall parameters like duration of rainy season, rainfall amount, number of rainy days and water surplus/deficit were determined for different weeks of onset of monsoon and presented in Table 3. The correlation between rainfall characteristics and onset of monsoon is also determined. The results reveal a negative correlation coefficient (r) for all the rainfall characteristics except water deficit. Thus, the delay in onset of monsoon reduces the duration of rainy season, rainfall amount, number of rainy days and water surplus while it increases the water deficit. The same phenomenon has been observed by Stewart (1990) in Kenya. Similarly, Agnihotri and Murti (2001) have also observed that delay in onset of monsoon reduces the monsoon season rainfall. The duration of rainy season is inversely correlated (r = -0.91) with onset of monsoon while water deficit has strongly positive correlation coefficient (r = 0.83) with onset of monsoon. However, water surplus is poorly correlated (r = -0.37) with onset of monsoon.

The results presented in Table 3 also reveal an increase of water deficit due to early and delay in onset (31st SMW) as compared to normal onset of monsoon (27th SMW). Thus, the early and late onset of monsoon results in more water deficit as compared to normal onset of monsoon. The results also reveals an decrease of 50.2 per cent in water surplus and a decrease of 55.8 per cent in water surplus due to late (31st SMW) and early onset (25th SMW) as compared to normal onset of monsoon in 27th SMW (390.7 mm).

The weekly moisture adequacy index was also determined for different weeks of onset of monsoon and presented in Fig.1, which reveal that a significant decrease in length of crop growing period as the week of onset of monsoon was delayed. The Fig. 1 also reveals that length of crop growing season reduces drastically if the onset of monsoon is excessively delayed (31st SMW). On the contrary excessively early onset of monsoon (25th SMW) results in maximum length of growing season except initial dry spells in general, while a decrease in length of growing season was observed due to delay in onset of monsoon (from 26th to 29th SMW) as depicted in figure.

Table 3 : Effect of onset of monsoon on normal water balance parameters

Onset of monsoon SMW	Rainfall amount mm	Duration of rainy season, week	Water surplus mm	Water deficit mm
25	457.5	12.0	172.6	124.7
26	694.8	10.8	475.5	79.5
27	578.6	11.6	390.7	95.8
28	552.8	11.8	374.3	107.2
29	469.1	7.1	344.5	50.2
31	322.0	5.0	194.6	78.4
Correlation Coefficient	-0.57	-0.91	-0.37	0.83

* Significant at 5% probability level, r = 0
Suggested crops and cropping system

Crop Planning Strategies

- Rainfall, climatic shift and drought analysis-Suggests for management of drought and crop planning at micro level (Tehsil) instead of district level
- Criterion for determination of onset of monsoon-Early onset of monsoon' requires more quantum of rainfall as compared to normal while 'delay in onset of monsoon' requires less quantum of rainfall
- 'Dynamic' crop planning concept- 'Select the crop, cultivars and cropping systems on the basis of time of onset of monsoon and available length of growing season and rainy days for that particular week of onset of monsoon and 'Onset of monsoon' approach is rational for analysis of rainfall data for crop planning

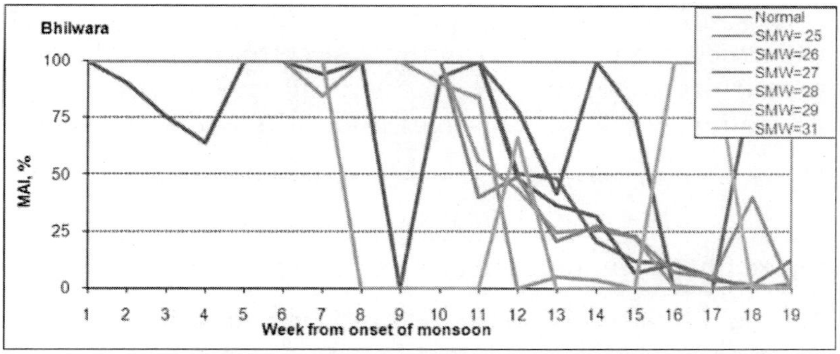

Figure 1: Moisture adequacy index for different weeks of onset of monsoon

Tehsil	Climate	Onset of monsoon		
		Early	Normal	Late
Asind Hurda Raipur	Arid	Maize	Maize + Black gram (2:2)	Sesame, pulses; sorghum, bajra, and cowpea fodder
Bhilwara Jahazpur Kotri Mandalgarh	Semi arid	Groundnut / maize + pigeon pea (1:1)	Groundnut + sesame (6:2), maize, groundnut	Maize + Black gram (2:2) , sorghum , pulses, sorghum fodder
Banera Mandal Sahara Shahpura	Arid to semi arid	Maize , Groundnut	Maize + Black gram (2:2) , Groundnut + sesame (6:2), maize	Maize + Black gram (2:2), sorghum fodder , sesame, pulses

- Water balance studies -Reveal that rainwater is surplus during early crop growth stages followed by deficit in later part, which suggests for more emphasis on rainwater management. in-situ rainwater management practices in arid tehsil while on ex-situ rainwater management practices in semi-arid tehsils . Dry spell analysis suggests preparedness to carryout farm operations more efficiently.

Conclusions

- The Morris and Zandstra method for determination of onset of monsoon has been evaluated and modified to develop a suitable criterion (AKON-4)
- An innovative approach for rainfall analysis 'Onset of monsoon approach ' has been suggested for crop planning which is more rational and practical as compared to conventional 'meteo-

rological approach' as it provides an opportunity to suggest dynamic crop planning (multiple set of cropping plans) based on different conditions of onset of monsoon for efficient utilization of natural resources under dryland conditions.

- Dynamic crop planning

  Early onset of monsoon - Long duration crops like groundnut, maize (medium duration) and its inter-cropping with pigeonpea

  Normal onset of monsoon - Maize (short duration), black gram, green gram, sesame and intercropping systems viz.; Maize +black gram and groundnut + sesame.

  Delayed onset of monsoon - Short duration kharif pulses (black gram and green gram and oil seed (sesame) in preference to long duration cereals.

References

1. Agnihotri, Y. and Murti ,Ram.2001. Relationship between onset of monsoon with sowing weeks and monthly/seasonal rainfall at Chandigarh. J. Indian Water Reso. Soc. 21(2): 60-64.
2. Allen, R.G., Pereira, I.S., Daes, D. and Smith, M. 1998. Crop evapotranspiration, Guideline for Computing Crop Water Requirements. FAO Irrig. and Drain, Paper 56. Rome, Italy. p. 300.
3. Deora, B.S. 2005. Modelling of Drought under different Agro-climatic Zones of Gujarat Maharashtra. Unpublished Ph. D. Thesis Deptt. of Soil and Water Engg. CTAE, MPUAT, Udaipur. pp. 264.
4. Jat,M.L., Singh, R.V., Balyan, J.K. and Jain, L.K. 2005. Rainfall analysis for crop planning in Udaipur region. Indian J. Soil Cons. 33(3): 264-266.
5. Morris, R.A. and Zandstra, H.G., 1979. Land and climatic in relation to cropping patterns. In rainfed low land rice, selected papers from 1970. Int. Rice Res. Conf. IRRI., pp. 255-274.
6. Pandarinath, 1991.Markov Chain model probability of dry and wet weeks during monsoon period over Andhra Pradesh.Mausam, 42(4): Pp. 393-400.
7. Panigrahi, B. and Panda, S.N. 2002. Dry spell probability by Markov Chain model and its application to crop planning. Indian J. Soil Cons. 30(1): 95-100.

Chapter 24

Contingent Crop Planning for Aberrant Weather Conditions

P.L. Maliwal

Emeritus Professor (ICAR) and former Director Research, MPUAT, Udaipur (Raj.)

Introduction

In arid and semi-arid tropics of India, rain is the only source of irrigation for sustaining crop grown under rainfed conditions. Rainfall in its quality and distribution is highly unpredictable. Aberrant weather appears in any one of the following ways:

i) Delayed onset of monsoon

ii) Dry spell immediately after sowing

iii) Short or long dry spell during the growing season in mid season or late season (break in monsoon–mid or late season drought)

iv) Early withdrawal of monsoon

Delayed onset of monsoon

Rain may be delayed by about 20-40 days as observed in some areas. Such a late onset shortens the length of growing season, restricts the choice of crops and varieties, limits use of inputs, results in low yields of *kharif* crops due to delayed sowing, and leaves less moisture for post rainy season crops grown in succession. The suggestions for delayed onset of monsoon are as under:

- Sowing under dry conditions
- Raising nursery and transplanting
- Changing plant population and planting geometry
- Selection of right crops and species depending on the onset of monsoon

Dry spell immediately after sowing

During some of the years, a dry spell may occur immediately after sowing of the crop. This may result in poor germination due to crust formation and poor establishment of crop stand due to inadequacy of soil moisture at sowing. The suggested measures for this weather condition are as under:

Re-sowing : When there is a severe dry spell and the population is too poor to expect any reasonable yield,resowing of the entire field using same or new crop using subsequent rains is necessary.

Gap filling : If the dry spell following sowing results in sub-optimum plant stand with the subsequent showers, gap filling may be considered advantageous. If the sown crop is suitable for transplanting and the stage of the seeding is adequate for transplanting in the gaps using excess seedling growing at some places in the field with subsequent rains. If the break in rains is rather long and do not permit transplantation of already sown crop with existing seedlings due to age, other crops grown in the region can also be used for transplanting e.g. pearlmillet, fingermillet, vegetables,etc. Similarly, reseeding can also be considered in the gaps using crops suited for late planting.

Break in monsoon–mid or late season drought

In dry lands, moisture is insufficient for crop production quite often. Farmers of dryland areas take up sowing when the soil profile is wet for 30-40 cm because it is sufficient to meet initial crop requirement of the moisture. Plants tolerate stress upto one month and later on start wilting. During this stage, there is need to introduce or superimpose mid season correction to save standing crop otherwise whole crop may be lost. Several practices have been developed and suggested to mitigate the damaging effects of dry spells as follows:

Thinning of the stands : Thinning of the plant stand is recommended to lessen the severity of moisture stress of about 30-50 days. Thinning of alternate plants in the rows in the event of mild stress and every third row under higher moisture stress condition is recommended. Thinning is resorted particularly to the most sensitive component for soil moisture, such as sorghum in sorghum + arhar intercropping system.

Removal of weak seedilings : In most of the dryland crops there may not be uniform emergence because of soil variation, differential depth of sowing or inadequacy of soil moisture at sowing. In depressions, there may be quick germination and on ridges there may be slow germination. In most of the cereals, the late germinated seedlings will not be vigorously growing and may not be competitive to give any yield. At severe drought conditions, these weak seedlings may be removed.

Repeated and deep intercultivation : Early breaks in rainfall brings about soil cracking early in the season which enhances evaporation loss of moisture through the cavities. To prevent soil cracking and to break the capillary loss, early and frequent inter-cultivations are needed. Generally blade hoes are employed for the purpose which penetrate hardly to a depth of 5 to 7.5 cm. Shovel type of inter-culturing hoes with iron coulters should be used which due to sharp cutting edge can tear off the soil and create a dust mulch to a depth of 10-12 cm.

Mulching : Mulches have a definite controlling effect on the evaporation of soil moisture and hence, play an important role in dryland agriculture. Under field conditions, surface soil is just stirred and the soil becomes air-dry. By adopting this practice, some amount of moisture is lost from the upper layer. After the preliminary loss of moisture, the soil mulch effectively controls further evaporation from deeper layers by breaking the capillary continuity. This system is highly suitable for medium to heavy textured soils.

Stripping of older leaves : Older leaves are shaded, they transpire more than helping in photosynthesis. So, to avoid moisture loss, these older leaves may be stripped, which helps in active growth of new parts. For example, in sorghum 12-15 leaves are produced, when the plants are at flag leaf stage strip off the bottom 3-4 leaves and use them for mulching or for feeding to the livestock. Similarly, in maize growing areas, it is a practice to cut the top portion above the cob after milky stage.

Returning ratooning : In case of pearlmillet and sorghum, when crop is exposed to dry spells of 30-35 days during peak growth (blooming stage), yields are drastically reduced. Under such conditions cut/harvest the crop for fodder purposes and allow the ratoon with the onset of rains (after experiencing drought). For quick re-growth supplemental addition of fertilizer is suggested.

Life saving irrigation : If there are farm ponds/seasonal streams, life saving irrigation should be given to save the crop. Runoff is inevitable in the tropical and sub-tropical climated of the country.

Use of anti-transpirants : The possibility of reducing transpiration is of greatest importance in dry farming. Any material that is applied to transpiring plant surfaces for reducing water loss from the plant is called anti-transpirant. Many herbicides fungicides and growth hormones posses anti-transpirant propertieis may be of stomatal closing type or film forming type or reflectant type or growth retardant. Some of the fungicides like phenyl mercuric acetate (PMA) and herbicides like atrazine in low concentrations serve as anti-transpirants by inducing stomatal closing. The plastic and waxy materials which form a thin film on the leaf surface retard the escape of water due to formation of a physical barrier e.g. mobleaf, hexadeconol, silicone. The success of these chemicals is limited since they also reduce photosynthesis. Some of the white materials form a coating on the leaves and increase the leaf reflectance (albedo). Transpiration loss is reduced by reflecting the radiation reducing leaf temperature and vapour pressure gradient. Spraying of kaolin (5%) forms thin layer on the leaf growth. Retardants such cycocel induce stomatal closure and thus, reduce transpiration losses from leaf surface .

Harvest the crop for fodder : If there is complete failure of rains and crop is not in position to produce seed yield, especially the *kharif* crops, in a double cropped area, such crop may be harvested for fodder e.g. sorghum, maize, bajra and the land could be kept ready to make use of shower for post rainy season crops like chickpea, mustard, etc.

Early withdrawal of monsoon : If monsoon terminates earlier, then it cuts the length of growing season. Normally crops in their maturity and those growing on shallow soils suffer heavily. In this condition, sowing of *rabi* crops may also get affected. It is, therefore, suggested to use short duration and early maturing varieties to face the situation in well-identified dryland areas.

Crop Group and Crop Wise Contingency Plan

i) Pearlmillet/ small millets

- Planting of pearlmillet hybrids is not advisable. Even early maturing hybrids like HHB 67 should not be planted beyond July.
- Wherever crop has been planted and is suffering for moisture stress, the plant population may be reduced and shallow inter-culture (dust mulching) may be practiced.

- The Pearlmillet may be grown mixed with pulses up 1st week of August as fodder crops, which may give atleast fodder to some extent.
- In years of extreme drought and erratic monsoon small millets like foxtailmillet for Rajasthan, Kodo, proso and little millet for Gujarat and little millet (Kutki) and Kodo millet for M.P. could be ideal crops for contingency planning for producing quickly fodder and grain.

ii) Sorghum

- In case there is delay in monsoon by 2-3 weeks, short duration cultivars, such as CSH 14 can be preferred.
- In case of staggered/delayed planting increase seed rate to an extent of 1.5 times and application of 20 kg of carbofuran or phorate (3g) granules in the seed rows before sowing to safeguard the crop against the anticipated shoot-fly attack is recommended.

iii) Upland rice

- In traditionally rice growing rain-fed areas, where rains are likely to come late and where a normal transplanted rice crop is ruled out, short duration upland rice varieties or those rice varieties that are suitable for direct seeding either in dry or wet condition and subsequent either in dry or wet condition and subsequent flooding are recommended by direct seeding.
- In certain areas, delayed transplanting can be done with older nurseries, if these varieties are suitable.

iv) Maize

- Planting of maize for grain after 15th July is not advisable.
- Inter-cropping of blackgram or pigeonpea with maize is better under dryland conditions.
- After 15th July, plant maize only for fodder purpose.

v) Pulses

- In the event of late monsoon arrival, the pulses crops which can be planted preferably upto 15th August or till 1st week of September are: blackgram, greengram, pigeonpea and mothbean.

- In case of greengram and blackgram all varieties recommended for *kharif* season may be grown upto first week of September. Sowing after first week of September may be done only for the purpose of fodder.
- Only specific varieties of pigeonpea may be grown upto first week September which are recommended for pre-*rabi* planting. These varieties are Bahar, Pusa-9 and *rabi* arhar 20(5) for the states of Bihar, West Bengal, Odisha. Andhra Pradesh, Maharashtra and Gujarat. Planting of early pigeonpea is not possible in states like Punjab, Haryana and western U.P., where early maturing varieties of pigeonpea normally grown in the first week of July.

vi) Oilseeds

- In case of oilseed crops, sesame can be planted upto third week of August, whereas niger can be sown upto September. Since the per hectare requirement of seed in case of sesame is very low, all out efforts must be made to cover more area under sesame to utilize the land in the event of failure of other *kharif* crops like maize, pearlmillet and other minor millets.
- Similarly, castor can be grown successfully upto first week of September in areas, which are suitable for castor cultivation.
- The other two important oilseed crops like soyabean and groundnut are likely to suffer heavily in case rains are delayed beyond July end. However, groundnut can be planted upto September in a few southern states. Therefore, efforts should be made to cover more area under groundnut in the states of Andhra Pradesh, Tamil Nadu and Karnataka.
- In delayed rainfall conditions, there seems good possibility for toria in the states of Punjab, Haryana, Madhya Pradesh, Uttar Pradesh , Bihar and West Bengal. All recommended varieties of toria can be promoted as a *rabi* crop.

vii) Cotton

- Use of micro-irrigation system, such as drip and sprinkler may be advocated whereever feasible and in the event of limited water availability.

- Development of ridge and furrow across the slope for effective conservation of soil moisture as well as rainwater.
- Use of organic mulches, such as subabul loppings, straw, etc. to conserve the soil moisture.
- Repeated intercultural operation to keep the field weed-free.
- Wherever water resources are available, such as lake, ponds, wells, etc. protective irrigations can be given to the crop.
- Micro - irrigation system, as suggested above, may also be adopted wherever possible for improving the water use efficiency and to cover more area.

viii) Vegetables

- Mulching soil surface with organic material (leaf mold or FYM) and clean cultivation.
- Growing vegetables, such as clusterbean, cowpea, lablab bean which can sustain with less amount of water.
- Enhancing cucurbitaceous vegetables by raising nursery in polythene bags followed by transplanting in order to save 2-3 irrigations.

Contingent Plan Relevant to All Crops and Cropping Systems:

i) Special package of practices

- Making fields free of weeds for full utilization of water and nutrients by the crops.
- Reduction of plant density in case of mid season drought, the crops should be suitably thinned out or ratooned in case of sorghum. In case drought occurs at very early stage, it is always better to resow with subsequent rain rather than allowing sub-optimal poor plant stand to persist.
- For drought conditions anticipating prolonged dry spell breaks the practices of inter-cropping helps in risk distribution. This can be achieved by including a companion crop like blackgram, greengram, cowpea with the main crops like maize, sorghum and pearlmillet.

- The recommended dose of nitrogen application should be reduced by 40 per cent underunirrigated conditions and should always be applied at sowing/planting.
- Full recommended dose of P and K should be placed as basal dose.
- Resorting to maximum possible use of organic bulky manures.
- Recommended plant protection measures to be adopted for *kharif* millets and pulses.
- Seed of sorghum, pegeonpea, greengram, cowpea ,groundnut, sunflower and castor must be treated with proper fungicides as per recommendation.

ii) Soil and water management practices

- Seed beds to be kept ready so as to facilitate sowing immediately with the onset of rains. Field should be properly levelled for uniform water distribution within the sub-plot.
- Broad bed and furrow, ridge and furrow, compartmental bunding and contour trench land configuration may be adopted in shallow Alfisols as moisture conservation practices.
- Furrow sowing of *kharif* crops at closure plant to plant distance with wider inter-row spacing.
- Frequent inter-culture to facilitate effect of loose soil as dust mulch. Wherever economically viable, mulching should be practiced in between crop rows using locally available mulching material.
- Wherever possible run-off may be harvested to make provision for protective irrigation at later stage/crop ripening.
- Major emphasis should be on in situ rainwater conservation, harvesting excess run-off for re-use and groundwater recharge.

References

1. Robertson, G.W. 1976. Dry and wet spell UNDP/FAO, Tun Razak Agric. Res., Center, Sungh: Tekam, Malaysia , project field report, Agrometeorology. A-6. p.-15.
2. Satpute, G.U. 2004. Characterization of Drought under different Agro-climatic Zones of Maharashtra. Unpublished Ph. D. Thesis Deptt. of Soil and Water Engg. CTAE, MPUAT, Udaipur. pp.199

3. Singh, P.K.2005. Rainfall variability and crop planning in Sabour region of Bihar, Journal of Agrometeorology 7(2). Pp 284-290.

4. Singh, J.B. and Hazra ,C.R.. 1999. Rainfall characteristics of Jhansi for crop planning under rainfed condition- A decadal change. Journal of Sil and Water Conservation, Vol. 43. pp 25-34.

5. Stewart, J. 1991. Principles and performance of response farming .pp 361-382.In: Climatic risks in crop productions- Models and management for semi-arid tropics and subtropics, Edited by Russcell C. Muchow and Jennifer A. Bellamy, CAB International, United Kingdom.p.548.

❑❑❑

Chapter 25

Renewable Energy Technologies for Environment and Biodiversity Protection

Deepak Sharma

Professor, Department of REE, CTAE, MPUAT, Udaipur

Introduction

Over the last four decades, there has been a complex challenge with Quality of environment, Health and Livelihood, Social Relationships, Economy, Technology and Politics. Question arises that Is this challenge only in the developing world or is the global phenomenon? Or is the cause Underdevelopment? In 1970, "Appropriate Technology" becomes popular and in 1987,"Sustainable Development" gains currency through proper management.

Sources of energy

Renewable which includes Solar energy, Hydro energy, Wind energy, Biomass energy and Geothermal energy.

Non-renewable includes Bituminous coal, Lignite, Oil, Gas and Radioactive elements.

WHO February, 2016 Report

It gave details that 3 billion people still depend on firewood, dung cakes and coal. Around 4.3 million people die every year. Mostly deaths occur due to pneumonia, stroke, heart disease and cancer. Black carbon and methane are climate changer.

Maximizing use of Renewable energy i.e. Solar, Wind, Bio gas, Gasification, Pyrolysis, Geothermal can reduce negative impact of climate change.

Role of Renewable Energy

Solar energy : The sun is an Ultimate Renewable Power Supplier. It's basic source of energy on the Earth. It has been estimated, that in one hour the Earth receives volume of energy from the Sun corresponding with the use of energy by the entire mankind in the whole year! However, it's not easy to use these resources.

Conditions favourable to acquiring of solar energy:

- Cloudless weather
- Long period of sunlight
- Low air dusting level

Figure 1: Fertilizer spraying using power generated by solar energy.

Figure 2: Applications of solar energy for water pump, street light and lamp lighting purpose.

NET METERING - Grid Solar Photovoltaic Power

It is a new concept - environment friendly and power efficient electricity evaluation system. Net metering can increase the energy in the power grid to keep up with demands during peak times, which is of particular interest to states facing power shortages. DC power generated from SPV panel is converted to AC power using power conditioning unit. It fed to the grid either of 33 kV/11 kV three phase lines or of 440/220 volt three/single phase line. Beneficiary pays to the utility on net meter reading basis.

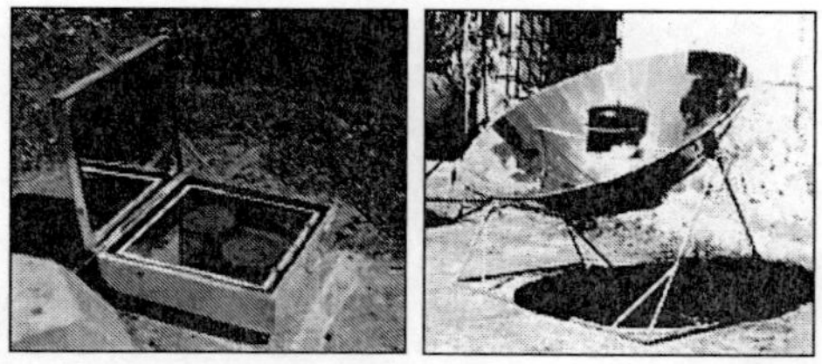

Figure 3: Emerging Solar Thermal Technologies -Box type and Concentrate type cooker.

Figure 4: Emerging Solar Thermal Technologies- Community type solar cooker.

Solar Water Heater

It is a flat plate collector type and Evacuated Tube collector type. It has a both Passive type and Active type systems. Temperature of hot water rises upto 80-90^{0}C. It has storage temperature of about 65°C.

Figure 5: Emerging Solar Thermal Technologies - Solar Water heater.

Solar Tunnel Dryer

It mainly uses for bulk drying chilies, fruits, peas, etc. Average temperature inside the tunnel found 15 20°C higher than the ambient temperature. Pay back period : 80-100 working days.

Figure 6: Emerging Solar Thermal Technologies - Solar tunnel Dryer

Wind energy

As far as 4000 years ago Babylonians used to pump water using wind turbines, when irrigating fields and drying wetlands. Well before then wind was being used in ship transport. From the 6th century Persians used to grind grain in windmills. Coastal areas, open plains, hills tops and mountain passes are the most suitable areas for acquiring wind energy. Wind energy doesn't cause air pollution (and in particular CO2) and doesn't contribute to depletion of reserves.

Energy from Biomass Waste

It is an anaerobic digestion of organic waste. Safe disposal of waste is possiblein bio gas technology. It is a matured technology. It has estimated potential of around 1.24 Crore of India. In Rajasthan, estimated potential is appro x 9.15 Lakh. Cumulative physical achievements as on 31-03-2017(Plants in Nos.) is 47.98 lakh (Approx. covered 39.98%). Target for year 2015-16 is around 1.11 lakh family type bio gas plants. It has Saved 2.79 lakh ton of fuel wood. It has production of approx 10617 tonnes of urea equivalent per annum and of 10,03,750 tonnes of organic manure production per annum. It has capability to prevent an estimated 33,880 tonnes of carbon dioxide emission into the atmosphere annually.

Figure 7: Bio gas Plant.

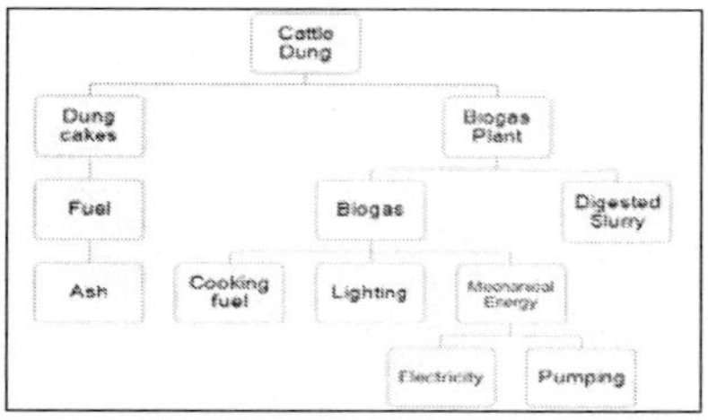

Figure 8: Flow Diagram of Bio gas Technology.

Bio gas generation, Purification and Bottling Development in India

It can become alternate for CNG, Waste Management. By-product of bio gas i.e. Bio Manure can be an additional source for income. Bio gas purification technology, such as water scrubbing using low or high pressure and pressure swing adsorption (PSA).

Briquettes

It is a compressed state of loose and dry biomass. Loose biomass has density of 200-300 kg per m^3, whereas Briquettes density has around 1200-1300 kg per m^3. It has low ash content (2-4%). It has high burning efficiency.

Figure 9: Flow diagram of Briquetting process.

Biomass Gasifiers

It converts organic or fossil fuel based carbonaceous materials into inflammable gas. It has use in thermal application & power generation. It has 40% inflammable gas like CO, H_2 and CH_4. It is mainly two types- Down draft & Up draft.

Improved Biomass Cook Stove

It burn fuel wood with controlled condition. It has lower wood consumption. It has complete smoke less process. It is also cheaper, as it has lower maintenance cost.

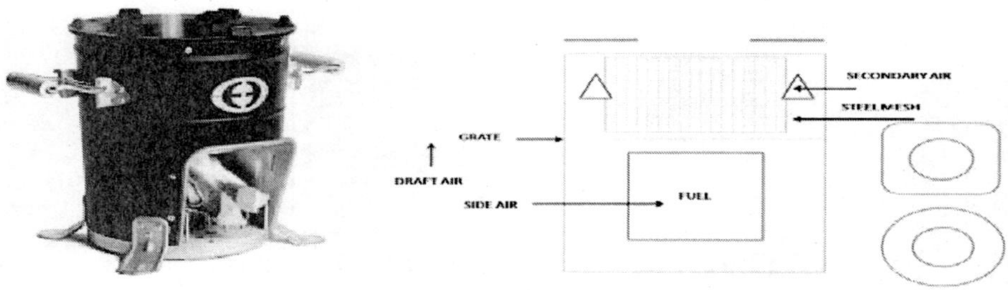

Figure 10: Diagram of Improved biomass cook stove.

Conclusion

Core drivers for Renewable Energy Technology is Energy Security i.e. 60% power generation based on coal, 23% import dependency of coal and 28% import dependency of oil. Electricity Shortage is main driver as increase in installed capacity by more than 110 times in 62 years but still there is a shortage of 43000 million units of electricity availability. There is planed and unplanned load shedding measure were required to be under taken utilities to bridge this demand supply gap. Energy Access is the other driver as around 85% rural household depends on solid fuel and 45% rural households are not able to access electricity. Last and most important driver for Renewable energy is Climate Change as India has taken a commitment of reducing emission intensity of its GDP by 20-25% from 2005 levels by 2020.

Chapter 26

Hi-tech Plant Propagation – A Methodology of Mitigating Stresses

Dhurendra Singh
Central Institute for Arid Horticulture, Bikaner-334006

Introduction

Arid environments are extremely diverse in terms of their landforms, soils, fauna, flora, water balances, and human activities. Because of this diversity, no practical definition of arid environments can be derived. However, the one binding element to all arid regions is aridity. Aridity is usually expressed as a function of rainfall and temperature. A useful "representation" of aridity is the following climatic aridity index: p/ETP, where P=precipitation; ETP=potential evapotranspiration, calculated by method of Penman, taking into account atmospheric humidity, solar radiation, and wind. Three arid zones can be delineated by this index: namely, hyper-arid, arid and semi-arid. Of the total land area of the world, the hyper-arid zone covers 4.2 per cent, the arid zone 14.6 per cent, and the semi-arid zone 12.2 per cent. Therefore, almost one-third of the total area of the world is arid land. The Indian hot arid region occurs between 22°30' and 32°05' N latitudes and from 68°05' to 75°45' E longitudes, covering western part of Rajasthan (19.6 m ha, 69%), north-western Gujarat (6.22 m ha, 21%) and south-western part of Haryana and Punjab (2.75 m ha, 10%). The majority of the hot arid zone comes under northern-western part of Rajasthan covering 12 districts.

Abiotic stress

The climatic elements, such as solar radiation, temperature, precipitation, humidity, evaporation, evapotranspiration and wind velocity affect growth and development of saplings. These climatic factors are not favourable for successful propagation of various horticultural crops in arid region. The main climatic features of Indian hot arid agro-ecosystem is high wind speed, sand storm and high temperature during summers upto 48-50^{0}C and during winter season temperature falls to the 2-3^{0}C, low atmospheric humidity i.e. 20-30 % during maximum period of the year. Moreover, the relative humidity goes down to below 10 per cent during summer months.

Biotic stress

During propagation of plants through tissue culture, green house or hydroponic system, young seedlings, rooted cuttings and grafted plants may get infection by various pathogens, such as Pthyium, Phytophthora, Fusarium, Cylindrocladium, Scelerotiania and Rhizoctonia solani insect–pests, birds and rodents can sometimes be devastating, especially under open condition and warm humid condition of greenhouses.

Facility of Hi-tech plant propagation

For mitigation of biotic and abiotic stresses, three important facilities of hi-tech propagation of fruit crops are required, one is greenhouses/protected structures for development of rootstocks during unfavourable conditions, adventitious root formation in cuttings and providing congenial environment for budding and grafting operations. Another facility of tissue culture laboratory and hydroponic system is also becoming an integral part of Hi-tech nurseries to improve the propagation efficiency and quality of saplings. Third is the hardening facility, under which young saplings are subjected for gradual acclimatization by passing them to different environmental regimes.

I. Green houses/protected structures facility

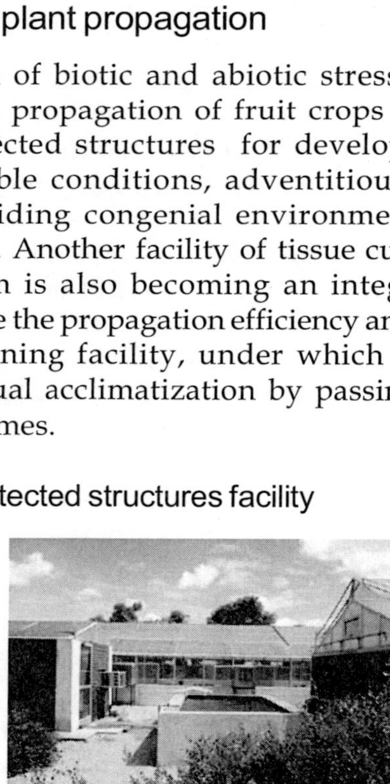

Hi-tech green house for plant propagation

Factors affecting propagation of fruit crops

Propagation of fruit crops affected by several factors, such as :

- Microclimatic conditions (light, water-relative humidity, temperatures and gases).
- Edaphic factors (propagation medium or soil, mineral nutrition and water).
- Biotic factors-interaction of propagules with other organisms (such as beneficial bacteria, mycorrhizal fungi, insect pests, pathogens, etc.).
- Physiological conditions of mother stock plant.
- Physiological conditions of scion plant.
- Hardening and post propagation care.

Methods of Hi-tech propagation

1. Development of seedling under protected conditions

Conventionally seedling of horticultural crops are produced in seed beds or in poly bags of different sizes, which require more and frequently watering. The modern system of seedling production in portrays as plug production has several advantages over the conventional system which are as following:

- The water requirement for seed germination and sapling growth is very less.
- This method is very useful in extending the growing season by producing seedling under protection and subsequently transplanted in either field (vegetable) or in polybags (fruit crops) as soon as the danger of frost, hails or low temperature is over.
- This method also avoid environmental hazards of germination of seeds by providing optimum germination conditions under low tunnel, poly trench, green or glass houses for good germination and uniformity of plants.
- No transplant sock occurs during field or container transplanting of plug seedlings.
- Faster growth of seedlings achieved for field production or budding onto rootstock.

- Seedling become rapidly ready for outdoor planting or transfer within four to six weeks.
- Protection from insect pests and diseases.
- Escape from environmental injuries.
- Initial irrigation for outdoor establishment is very low.
- Allow precision crop scheduling and easy in post nursery handling and transplanting.

2. Plug production technique

Under this technique large number of vegetable seedlings and rootstocks of fruit crops can be produced annually in low cost poly houses, environmentally controlled greenhouses and under poly trenches by controlling optimum environmental conditions for seed germination and growth of the seedlings. A plug is a sapling which is produced in a small volume of medium in small individual cell or group of cells of different sizes ranging from 50 to 250 cc.

Plug production of ber and khejri

3. Green house based multitier hi-tech propagation in arid zone

Looking into inherent environmental problems of arid ecosystem we have designed and established hi- tech propagation facilities, such as closed environmental chamber based on evaporative cooling system, glass house fitted with environmental controlled devices, such as fan and pad cooling system, micro irrigation system, foggers, misters and heating system and shade house fitted with cooling pad made of local available biomass of wild ber (Ziziphus nummularia) at CIAH, Bikaner. This multilayer approach was found effective in improving vegetative propagation efficiency of more than 15 species of arid horticulture crops and acclimatization of tissue culture plants of Aloevera and Cactus pear.

4. Hydroponic propagation

Preliminary studies conducted at CIAH, Bikaner under hydroponic system for nursery raising of high value crops like broccoli and lettuce. The system designed based on soilless culture media based on vermiculite fortified by micro and macro salts of MS medium. The water was applied through motorised pump and drained water collected and re-circulated for irrigating nursery bed. The seedlings get ready for field plantation within 03 weeks. Adventitious root formation in cutting of sweet potato and basella has been obtained in 2-10 days respectively under oxygenated hydroponic system kept under greenhouse.

Adventitious root formation Soilless nursery raising of lettuce and broccoli in stem cutting of Basella and sweet potato.

5. Mcropropagation

Micropropagation has become an important part of commercial propagation for many horticultural crops with following advantages:

1. Mass propagation of specific clones/variety/hybrid.

a) Plants whose natural rate of multiplication is relatively low for example date palm, banana, ker, orchids, bulbous vegetables and ornamentals, ferns gerbera, carnation and many foliage plants.

b) New cultivars high in market demands for getting market opportunity quickly.

c) Cultivars whose high value makes micropropagation a viable alternative to conventional methods. For example, date palm, banana, orchids and many foliage plants

d) Propagation to difficult- to- root. For example, many fruit crops, root stock genotypes.

e) Conservation of endangered species. For example, elite genotypes of fruits, vegetables and flowers.

2. Production of disease free plants

a) Elimination of virus, for example, Citrus, Chillies, Tomato and Potato.

b) Elimination of disease e.g. Banana, Citrus, guava, Papaya, etc.

c) Maintain germplasm and resource material in a pathogen free condition

d) Include micropropagation as the initial step in production for foundation clone system

e) Facilitate the distribution of the commercial material in the international trade

3. Clonal propagation of parental stocks for hybrid seed production

This procedure is used for hybrid seed production of tomato, cucurbits, marigolds, asparagus and broccoli.

- Maintenance of cytoplasmic male sterile lines
- Year round nursery production
- Germplasm conservation

Types of systems for regeneration of plantlets by micropropagation

Micropropagation is defined as the true-to-type propagation of selected genotypes using in vitro techniques. Four basic methods are used to propagate plants in vitro depending on the species and cultural condition, micropropagation can be achieved by:

1. Enhanced axillary shoot proliferation (shoot culture): Suitable explants

- Shoot tip
- Meristem tip
- Nodal culture
- Stool shoot
- Proliferation of pseudocorms
- Minitubers

2. Adventitious shoot formation: Suitable explant

- Internode section
- Leaf pieces
- Fragmented shoot apices
- Immature inflorescence
- Bulb scale

3. Micrografting

- Shoot/meristem tip

4. Non-zygotic embryogenesis : Suitable explant

- Cell
- NucellusTissue
- Callus
- Shoot tip

Developmental stages in micropropagation

Stage 0: Donor plant selection and preparation

Explant quality and subsequent responsiveness in vitro are significantly influenced by the phytosanitary and physiological conditions of the donor plant. Prior to culture establishment, careful attention is given to the selection and maintenance of the stock plants used as the source of explants. Numerous practices are employed to increase explant responsiveness by modifying the physiological status of the stock plant. These practices include the following: (1) trimming to stimulate lateral shoot growth; (2) pre-treatment sprays containing cytokines or gibberellic acid; and (3) use of forcing solution containing 2% sucrose and 200 mg/L 8- hydroxyquinoline citrate for induction of bud break and delivery of growth regulators to target explant tissues (Read, 1988).

Stage 1: Establishment of aseptic cultures

Initiation and aseptic establishment of pathogen eradicated and responsive terminal or lateral shoot meristems explants are the goal of this stage. The primary explants obtained from the stock plants may consist of surface disinfested shoot apical meristems or meristem tips for pathogen elimination or shoot tips from terminal or lateral buds. The presence of microbial contaminants can adversely affect shoot survival, growth, and multiplication rate. Bacteria and fungal contaminants often persist within cultured tissues that visually appear contaminant free. Consequently, it is essential that stage I cultures be indexed (screened) for the presence of internal microbial contaminants prior to serving as sources of shoot tip and nodal explants for stage II multiplication.

The following factors may affect successful stage I establishment of meristem explants:

1. Explantation time
2. Position of the explant on the stem
3. Explant size
4. Polyphenol oxidation.

Time of explantation can significantly affect explant response in-vitro. In deciduous woody perennials, shoot tip explants collected at various times during the spring growth flush may vary in their ability for shoot proliferation. Shoot tips collected during or at the end of the period of most rapid shoot elongation exhibited weak proliferation potential. Explants collected before or after this period is capable of strong shoot proliferation in-vitro (Brand, 1993). Often similar lateral meristem explants from the top and bottom of a single shoot may respond differently in-vitro in woody plants exhibiting phasic development, juvenile explants typically are often more responsive than those obtained form the often non responsive mature tissues of the same plant. Sources of juvenile explants include the following:

1. Root suckers
2. Basal parts of mature trees
3. Stump sprouts
4. Lateral shoots produced on heavily pruned plants.

The excision of primary explants often promotes the release of polyphenols and stimulates polyphenol oxidase activity within the damaged tissues. Oxididzed polyphenol products often blacken the explant tissue and medium. Accumulation of these products can eventually kill the explants. Procedures used to decrease tissue browning include :

1. Use of liquid medium with frequent transfer
2. Adding antioxidants, such as ascorbic acid or polyvinyl pyrrolidone (PVP)
3. Culture in reduced light or darkness.
4. Season of explantation.

- Knowledge of the specific sites of hormone biosynthesis in intact plants provides insight into the relationship between explant size and dependence on exogenous growth regulators in the medium. Endogenous cytokinins and auxins are synthesized primarily in root tips and leaf primordial, respectively. Consequently, smaller explants, especially cultured apical meristem domes, exhibit greater dependence on medium supplementation with exogenous cytokinin and auxin for maximum shoot survival and development (Shabde and Murashige, 1977).

A commonly held misconception is that primary explants exhibit immediate and predictable growth responses following inoculation. For many species, particularly herbaceous and woody perennials, consistency in growth rate and shoot multiplication is achieved only after multiple subculture on Stage I medium. Physiological stabilization may require from 3 to 24 months and four to six subcultures. Failure to allow culture stabilization, before transfer to a Stage II medium containing higher cytokinin levels, may result in diminished shoot multiplication rates or production of undesirable basal callus and adventitious shoots multiplication rates or production of undesirable basal callus and adventitious shoots. With some species, the time required for stabilization can be reduced by initial culture in liquid medium.

Stage II: Proliferation of axillary shoots

Stage II propagation is characterized by repeated enhanced formation of axillary shoots from shoot tips or lateral buds cultured on a medium supplemented with a relatively higher cytokinin level to disrupt apical dominance of the shoot tip. A subculture interval of 4 weeks with a three to eightfold increase in shoot number is common for many crops propagated by shoot culture. Stage II cultures are routinely subdivided into smaller cluster, individual shoot tips, or nodal segments that serve as propagules for further proliferation. Additionally, axillary shoot clusters may be harvested as individual un rooted Stage II microcuttings or multiple shoot clusters for ex-vitro rooting and acclimatization. Clearly, Stage II represents one of the most costly stages in the production process.

Both source and orientation of explants can affect Stage II axillary shoot proliferation. Subcultures inoculated with exhibit higher multiplication rates than lateral bud explants. Inverting shoot explants oriented explants per culture period in some species.

Selection of stage II cytokinin type and concentration is made on the basis of shoot multiplication rate, shoot length, and frequency of genetic variation. Although shoot proliferation is enhanced at higher cytokinin concentrations, the shoots produced are usually smaller and may exhibit symptoms of hyperhydricity depending on the species, exogenous auxins may or may not enhance cytokinin-induced axillary shoot proliferation. Addition of auxin in the medium often mitigates the inhibitory effect of cytokinin on shoot elongation, thus increasing the number of usable shoots of sufficient length for rooting. This benefit must be weighed against the increase chance of callus formation.

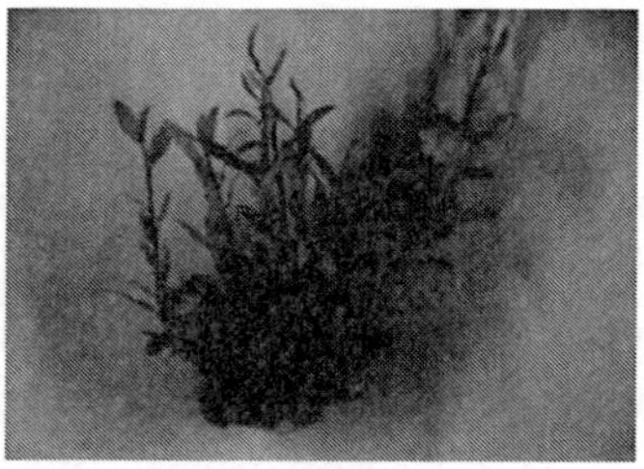

Profuse auxillary branching in ker

Stage III: Pre-transplant (rooting)

This step is characterized by preparation of Stage II shoots or shoots clusters for successful transfer to soil. The process may involve:

1. Elongation of shoots prior to rooting
2. Rooting of individual shoots or shoot clumps
3. Fulfilling dormancy requirements of storage organs by cold treatment
4. Pre-hardening cultures to increase survival.

Where possible commercial laborites have developed procedures to transfer Stage II microcuttings to soil, thus bypassing Stage III rooting. There are several reasons for eliminating Stage III rooting. Estimated costs for Stage III range from 35 to 75% of the total production costs. This reflects the significant input of labour and supples required to complete Stage III rooting. Considerable cost saving can be realized, if Stage III is eliminated.

For various reasons, however, it may not always be feasible to transplant Stage II microcutting directly to soil. Given the aforementioned limitations of Stage III rooting, Debergh and Maene (1981) proposed using Stage III solely to elongate Stage II shoot clusters prior to separation and rooting ex-vitro. Elongated shoots may be further pre-treated in an aqueous auxin solution prior to transplanting. Usually, Stage III rooting of herbaceous plants can be achieved on medium in the absence of auxins. However, with many woody species, the addition of an auxin (IBA or NAA) in Stage III medium is required to enhance adventitious rooting.

Stage IV: Hardening of plantlets

The ultimate success of micro propagation depends on the ability to transfer and re-establish vigorously growing plants from in-vitro to greenhouse conditions. This involves acclimatizing or hardening–off plantlets to conditions of lower humidity and higher light intensity. Even when acclimatization methods are carefully followed, poor survival rates are achieved. Micro propagated plants are difficult to transfer in-vitro condition due to following reasons:

1. Heterotrophic mode of nutrition
2. Poor control of water loss

During acclimatization, there is a need for plants to rapidly transition from the heterotrophic state for survival (Preece and Sutter,1991). Unfortunately, this transition is not immediate. For example, in cauliflower, no net increase in CO_2 uptake is achieved until 14 days after transplantation. This occurs only following development of new leaves since the leaves produced in-vitro in the presence of sucrose never develop. Photosynthetic competency interestingly, before senescencing, these older leaves function as "lifeboats" by supplying stored carbohydrate to the developing and photosynthetically competent new leaves. This is not the rule with all micropropagated plants since the leaves of some species become photosynthetic and persist after acclimatization.

These features include reductions in leaf epicuticular wax, poorly differentiate mesophyll, abnormal stomata function, and poor vascular connection between shoots and roots. To overcome these limitations, plantlets are transplanted into a well-drained "sterile" growing medium and maintained initially at high relative humidity and reduced light (40 to 160μmol. $M^{-2}s^{-1}$) at 20 to 27^{0}C. Relative humidity may be maintained with humidity tents, single tray propagation domes, intermittent misting, or fog systems. However, use of intermittent mist often results in slow plantlet growth following water-logging of the medium and excessive leaching of nutrients. Transplants and acclimatized by gradually lowering the relative humidity over a 1 to 4 week period. Plants are gradually moved to higher light intensities to promote vigorous growth. For successful acclimatization of plantlets, the following infrastructural facilities are required:

1. Pots of different sizes
2. Poting mixtures, such as vermiculite, perlite, cocopeat and vermicompost
3. Hardening chamber fitted with temperature, photoperiod and humidity control devices
4. Poly house/green house facility with controlled temperature and humidity factors
5. Shade house/Net house with fogging/misting devices
6. Shade house for further growth and development of plantlets in containers filled with a mixture of sand, soil and compost.

References

1. Aitken-Christie, J., T. Kozai and S. Takayama (1995) Automation in tissue culture - general introduction and overview. In : Automation and Environmental Control in Plant Tissue Culture. (Eds. J. Aitken-Christie, T. Kozai, and M.A.L. Smith), Kluwer Academic Publishers, Dordrecht, pp 1-18.
2. Apel, K. & Hirt, H. (2004). Reactive oxygen species: metabolism, oxidative stress, and signaltransduction, Annu. Rev. Plant Biol. 55, pp. 373–399.

3. Anonymous (2002) Annual Report, Central Institute for Arid Horticulture, Bikaner. Asada, K. (1999). The water-water cycle in chloroplasts: scavenging of active oxygens and dissipation of excess photons. Annu Rev Plant Physiol Plant Mol Biol 50: 601–639.

4. Bafeel, S.O. & Ibrahim M.M. (2008).Activities of antioxidant enzymes and accumulation of α-tocopherol in Medicago sativa seedlings subjected to dark and chilling conditions, Int. J. Agric. Biol. 10(6): 593-598.

5. Biswal, U.C., Biswal, B. & Raval, M.K. (2003). Chloroplast Biogenesis: From Proplastid to Gerontoplast, Springer, Dordrecht, The Netherlands.

6. Biswal, B.(2005). Photosynthetic response of green plants to environmental stress: Inhibition of photosynthesis and adaptational mechanisms. In Handbook of Photosynthesis (ed. Pessarakli, M.) CRC Press, Florida, USA, 2nd edn, pp. 739–749.

7. Biswal, B.; Joshi, P.N.; Raval, M.K. and Biswal, U.C. (2011). Photosynthesis, a global sensor of environmental stress in green plants: stress signalling and adaptation. Current Science, Vol. 100(12):25.

❑❑❑

Chapter 27

Value Addition of Agriculture Produce

Prakash Panwar

Assoc. Professor, College of Agriculture, Bhilwara

Introduction

Value addition is a process in which for the same volume of a primary product, a high price is realized by means of processing, packing, upgrading the quality or other such methods. Value added agriculture is a process of increasing the economic value and consumer appeal of an agricultural commodity. It is a production/marketing strategy driven by customer needs and preferences. Examples of value added agriculture are: food processing, (including: drying, canning, juicing, unique packaging, labelling and marketing.

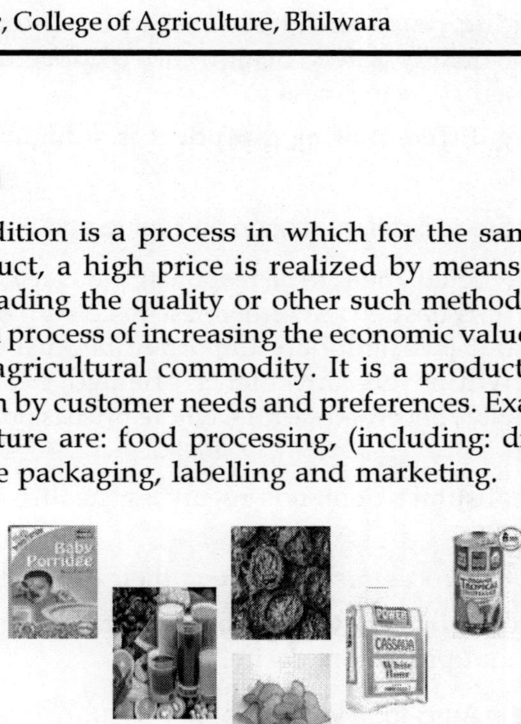

Need for value addition

Value addition improve the profitability of farmers, empower the farmers and other weaker sections of society, especially women through gainful employment opportunities and revitalize rural communities. It provides better quality, safe and branded food to the consumers. It emphasizes primary and secondary processing. It reduces post harvest losses. It helps in reduction of import and meeting export demands. It has way of increased foreign exchange. It encourages growth of subsidiary industries. It reduces the economic risk of marketing. It increases opportunities for smaller farms and companies through the development of markets. It diversifies the economic base of rural communities. It increases farmers' financial stability.

Why is it important to add value in the agro food value chain

- **Make more money**: an added value agricultural product has more market value than a commodity
- **Meet changing tastes and preferences** of consumers, such as: convenience, quality, safety, health, variety, price, and social and environmental consciousness
- **Compete by differentiating a product** in a highly competitive market

Market forces for product differentiation and value addition:

Producers have challenges to be responsive to consumer demands by producing what is desired. Attentiveness to consumer demands in quality, variety and packaging are important because demographic trends show growth in the convenience-oriented, health conscious and environmentally concerned sectors, where price is not as important as quality.

- Increased consumer demands regarding health, nutrition and convenience.
- Effort by food processors to improve their productivity.
- Technological advances that enable producers to produce what consumers and processors desire.

How to add value in Agro-food value chain

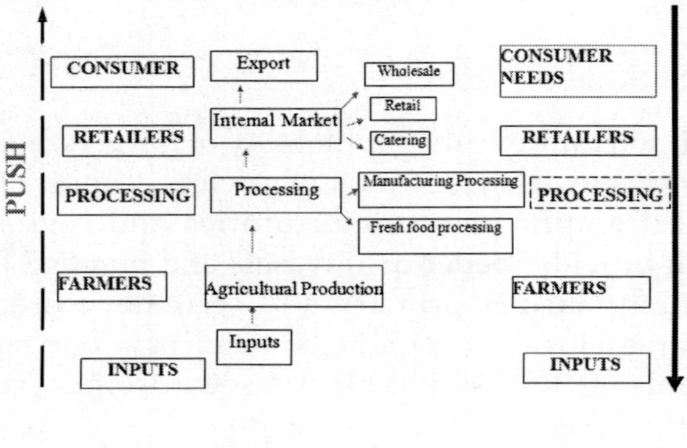

Cereals

Cereals supply the bulk of the food consumed by the human race. They are the cheapest source of food energy and constitute a high percentage of the calorie and protein intake of man, particularly in the developing countries. Cereal grains belong to the monocoty ledonous family, Gramineae or grass family. The principal cereal crops are rice, wheat, maize (corn), sorghum (jowar), millets, barley and oats. Of these cereals, rice and wheat are the chief cereals of human diet. However, as a source of carbohydrate related to landuse, maize ranks first among the cereal grains, followed by rice, sorghum and wheat.

Value added products of Sorghum

Value added products of Pearl Millet

Value added products of Maize

Horse Gram(Kulthi)-A super food in its own right

This pulse crop is widely cultivated in India since ancient times, and native to the South-East Asian subcontinent and tropical Africa. The US National Academy of Sciences has identified this Legume as a potential food source for the future,thanks to its exceptional nutrition profile, drought-resistance and general hardiness. Horse gram is the most protein-rich lentil found on the planet. It is very high-powered. That's why, race horses are fed with this gram, which is called horse gram in the market.

Wonderful Qualities of Horse Gram

- High in iron, calcium & protein. In fact, horse gram has the highest calcium content among pulses and is one of the richest vegetarian sources of protein.
- Low in fat and high in carbohydrate content.
- Low in lipid and sodium content & its slow digestible starch make ideal for diabetic & obesity patients.
- Ability to generate heat and energy in your system and therefore keeps you warm on a cold winter day.

AONLA-Indian goose berry

One of the most important indigenous fruit of Indian origin and is the richest source of Vitamin c at a very low price.

Value added products of Aonla

Conclusion

Value added agriculture increased the economic value and consumer appeal of an agricultural commodity. It is a production/ marketing strategy driven by customer needs and preferences.

❐❐❐

Chapter 28

Value Addition in Underutilized Fruits of Arid and Semi-arid Regions

R.A. Kaushik
Professor, MPUAT, Udaipur

Introduction

Fruit trees like ber, aonla, jamun, tamarind, custard apple, bael, lasora, phalsa, ker, pilu, etc. well-suited for arid environment, remained under exploited owing particularly to lack of appreciation of their quality as fruits, their economic potential and above all due to non-availability of the needed technology for their profitable cultivation. Because of their curative properties, these fruits have been used in Indian system of medicine, such as Ayurvedic and Unani since time immemorial. Apart from their nutritive and medicinal values quite a few of these underutilized fruits have excellent flavour and very attractive colour. In spite of these quality attributes, most have not undergone any conscious phase of domestication and human selection. Their cultivation is very restricted and they grow mainly wild. Being tolerant to biotic and abiotic stresses, these fruit species are suitable for growing in the disaster- and drought-prone areas.India has a rich heritage of indigenous fruit types.Significant advances have been made in recent times towards providing technology support for their cultivation.

The reasons for poor popularity of underutilized fruit crops are :

1. Lack of awareness about the economic benefits;
2. Non-availability of good quality planting materials;
3. Lack of technology to reduce the gestation period and enhance the fruit production;
4. Poor marketing network;
5. Lack of technology for value addition, through processing.

In this article an attempt is being made to summarise nutritional value, processing and value addition in underutilized fruits particularly from dry tract of India.

Ber (*Zizyphus mauritiana*)

The ber is found growing wild throughout the country. It is grown in abundance in plains of the Punjab, Haryana, Rajasthan, Gujarat and Uttar Pradesh. Ber grows practically in all types of soil. It withstands worst frost, prolonged drought and water-logging. It thrives well in comparatively dry regions. The wild ber tree is bushy and thickly set with thorns, which are much more numerous than in the cultivated variety. Wasteland areas can be covered with this fruit. Ber is very nutritive fruit. It is a rich source of vitamins and contains appreciable amounts of minerals. Ber gives much higher vitamin C than orange juice in equivalent quantities. Ber fruits are a blood purifier and help in digestion. They are useful in fevers, wounds and ulcer. Dried fruit is a mild laxative and expectorant, removes blood impurities, allays thirst. Seeds are given with buttermilk in bilious disorders. This fruit is not being processed at commercial scale, but it has lot of potential in dehydration and beverage industries to harness its benefits during off season.

Aonla (*Emblica officinalis*)

The aonla fruit is highly nutritive and one of the richest sources of vitamin C. The tree is very hardy and can be grown successfully under arid and semi-arid conditions. Aonla may be an important fruit of the future due to its high medicinal, nutritional value, high productivity per unit area and suitability even in the wastelands. Aonla is one of the important medicinal plants in India and the fruit is used as one of the ingredients in several Aurvedic medicines like Chayvanparash, Triphla, Amlaki, etc. The fruit is acidic, cooling, refrigerent, diuratic and laxative. It is useful in diarrhea, dysentery, diabetes, jaundice, dyspepsia and cough. Several value added products are being prepared from aonla and many of them are commercially manufactured by big companies too.

Bael (*Aegle marmelos Corr.*)

Bael is very hardy and has wider adaptability to adverse soil and climatic conditions. Bael tree thrives in soil having sodicity upto 30%

exchangeable sodium as salinity 9 dsm^{-1} ECE and can be grown upto an altitude of 1200 metres. It can also be grown on wastelands. Generally propagated through seed, bael can also be propagated through patch budding during June-July. Bael is known for its high medicinal and nutritional properties. Almost every part of the bael tree is used. The fruit is very rich in sugar, riboflavin (vitamin B_2) and minerals. The ripe fruit is a tonic, restorative, laxative and good for heart and brain. The mature fruit is astringent, digestive and stomachic and usually prescribed for diarrhoea and dysentery. Bael fruit is regarded as a valuable and efficacious remedy in diarrhoea, dysentery, chronic constipation and irregularity of the bowl.

Bael peel and pomace are the two processing wastes of bael, which are left over in abundance after the extraction of pulp. Keeping this fact in view, studies on the utilization of processing waste were carried out. Analysis of the wastes revealed that these are a rich source of protein, fiber and mineral like Ca and P. Hence, these can be used as animal feed. Palatability of the waste was adjudged by feeding the material to the experimental animals. It was observed that after mixing with cattle feed in 1:1, the wastes were readily accepted by the animals. In-vitro digestibility studies indicated that 69.7% of the pomace and 40.1% of the peel was digestible. Work done at HAU, Hisar on digestibility studies through actual feeding trails have shown upto 30% replacement of concentrate.

Phalsa (*Grewia subinaequlis*)

Phalsa is a small bush, which bears many small berries like fruits. It is one of the hardiest fruits, is drought resistant and requires little care. Phalsa is grown mostly from seed, but propagation from cutting with the use of plant growth regulators is also possible. It has immense potentialities for commercial cultivation, provided its use is exploited fully. Ripe fruits are sub-acidic and good source of vitamin A and C. They are fair source of phosphorus and iron. In early Vedic literature phalsa is mentioned for its medicinal values. The fruits are stringent and said to have a cooling effect and stomachic. It cures inflammation, blood disorders, fever and constipation becoming a digestive tonic. In Unani medicines, it is considered to be useful for heart, diarrhoea and fevers. It is also helpful in removal of dead foetus and is considered good for throat troubles. This fruit is a good source of anthocynins and can be exploite commercially for this purpose, beside preparing good quality beverages.

Karonda (*Carissa carandas*)

Karonda is an evergreen spiny shrub. It is very hardy and can grow in tropical and sub-tropical climatic conditions. Karonda can be grown on poor soils of arid regions. It is commonly propagated by seeds and vegetative methods like stem cuttings, inarching and air layering. Karonda fruit is a rich source of iron. The fruit possesses antiscorbutic properties. The fruit is useful in brain diseases. It also cures biliousness and blood impurities. Ripe fruit is cooling, appetiser and antipyretic. Picles and candy are value added products of Karonda.

Pilu (*Salvadora oleoides Decae*)

Pilu is a small, evergreen tree or shrub found in the arid regions of northern and western India. This indigenous tree is grown on dry and desert areas of Rajasthan, Haryana, Punjab, Gujarat, Uttar Pradesh and Madhya Pradesh. It can tolerate salinity to a great extent. The tree is suitable for growing as shelter belts and windbreaks. It grows wild in arid and sandy ares. The plant coppices fairly well and regenerates through root suckers and to some extent through natural layers. Seeds can also be used for regeneration. Pilu bears small, sweet, edible fruits. Fruit contains sugar, fat, colouring matter and alkaloids. The Fruit has a sharp, pungent, acrid and sweet, sour taste with a flavour, appetizer, laxative, carminative, alexipharmic, useful in piles, tumours, bronchitis diseases of spleen and ascites. Pilu pulp can be used for preparing several types of drinks, whereas its seed can be exploited for non-edible oil.

Ker (*Capparis decidua*)

Ker (teent) is a perennial woody plant, which grows wild on barren soils. In India, its main distribution is in Western Rajasthan, Punjab, Haryana, Central India, Gujarat and Deccan. It grows almost in all types of wastelands in arid zone soils. It can do well even in regions having rainfall as low as 150 mm. No attempts have ever been made for its scientific cultivation, although it may be second to none as far as its capability to stabilise sand dunes and combat drought with fair tolerance to salinity and alkalinity is concerned. Ker have immense potential as a source of nutrients for the inhabitants of areas prone to extreme drought and hot conditions, where other major food crops cannot be grown economically. Unripe fruit of Ker contained 14.88% protein, 7.43% fat, 5.96% total mineral matter, 12.32% crude fibre and 59.41% digestible carbohydrates on dry matter basis. The fruit has a sharp hot taste astringent to the bowls, destroys foul breath, biliousness and urinary purulent discharges good in cardiac troubles. Unripe fruit of Ker after processing are eaten as vegetable and pickle.

Custard apple (*Annona suqmosa*)

Sitaphal is a medium sized fruit which comes up well in arid, semi-arid and low fertility soils clubbed with dry climates particularly in Rajasthan, Northern Gujarat, dry parts of Uttar Pradesh, etc. The fruit is used for table purpose, pulp is used in ice creams, puddings, bakery items and confectionery. An alcohol 'ancorin' is extracted from custard apple and is considered to possess insecticidal properties. Due to presence of annonine, the leaves, stem and other portions of the plant are also bitter and the plant is not attacked by the goats or cattle. This is one of the main reasons for its spontaneous spread in nature. The root is a drastic purgative i.e. a cleansing agent of the bowels. Traces of hydrocyanic acid have been found in the leaves, bark, root and seeds. It has got medicinal properties too. Oil is extracted from custard apple seeds for making soap and the cake is used as manure.

At MPUAT, Udaipur a complete value chain has been developed. Pulp extraction technology standardized and commercialized. Seed used for extraction of anninine and peel converted into manure.

Jamun (*Syzygium cuminii*)

The jamun tree, native to India, thrives easily in hardy tropical regions and is found in all parts of our subcontinent as well as countries of South-east Asia and Eastern Africa. The fruit is acidic and astringent in nature, with a sweet taste. Due to its acidic nature, it is usually eaten with a sprinkling of salt.A medium-sized jamun provides about 3-4 calories. The seed of the fruit is also rich in protein carbohydrates and traces of calcium. It is also a source of iron, providing 1-2 mg per 100 grams and also vitamin C, providing 18 mg per 100 grams (the daily adult requirement of this vitamin is 40 mg). It provides folate and vitamin B, carotene, phytochemicals (anti-oxidants), magnesium, potassium and fibre. Jamun is believed to be of special use in treatment of diabetes. In Unani and Ayurvedic system, it is used to treat digestive disorders including diarrohea. Extracts of the bark, seeds and leaves have been found to cause a marked prolonged decrease in blood sugar and glycouria (sugar in urine). Several studies provide evidence that jamun has hypoglycemic effects with upto 30 per cent reduction in blood sugar reported in some studies. Seeds are rich in alkaloids which have hypoglycemic effects.

Tamarind (*Tamarindus indica*)

Tamarind has been introduced to India several centuries ago and is now extensively used in Indian cooking. However, tamarind has not been cultivated as a regular crop. The trees are established through natural seeding on community lands and farmers generally establish a few trees in their backyards or field bunds for household use. During the last 2-3 decades, the demand for tamarind has increased significantly because of the use of modern technologies for processing the pulp into ready-to-use cubes and powder.

Mahua (*Madhuca indica*)

Mahua is an oilseed tree which grows naturally on village common lands in Central India. Its flower consists of a sweet and juicy corolla which drops a few days after flowering. The local people, particularly the tribals, collect the fallen corolla and consume it in the fresh or dehydrated form. The seeds contain 30% edible oil. Long gestation period and poor income are major constraints for cultivation of mahua and many other underutilised tree crops. Techniques have now been developed to produce grafted plants of mahua, which start bearing in the third year. Looking to the economics and short gestation period, farmers are now keen to establish plantation of mahua on their degraded lands as a commercial crop.

Mahua flower juice based RTS blended with ginger and fennel significantly improved the organoleptic acceptance of the beverage. Higher level of blending was needed in case of ginger than that of fennel. The shelf-life of mahua RTS was five months under refrigerated conditions (Mishra *et al.,* 2013).

Table 1: Major processed products which can be prepared from underutilized fruits

Processed Product	Fruits
Jam	Jamun, Karonda, Aonla, Mulberry, Soursop, Tamarind, Wood apple
Jelly	Tamarind, Jamun, Karonda, Tamarind
Preserved	Ber, Aonla, Ker, Sangri,.Karonda, Bael , Karonda, Soursop
Candy	Aonla, Karonda, Tamarind
Glazed fruits	Tamarind, Annanas, Aonla
Confectionary	Amra, Aonla, Tamarind
Juice/Syrup/ Beverage /	Aonla, Ber, Bael, Jamun, Karonda, Phalsa, Mulberry, Pomegranate, Soursop, Wood apple, Tamarind
Wine	Mahua, Jujube, Ber, Indian fig, Karonda
Chutney	Karonda, Woodapple, Aonla
Sauce	Karonda, Tamarind, Woodapple, Pomegranate
Pickle	Jujube, Tamarind, Ker, Lasora, Gonda
Dehydration	Aonla, Karonda, Ker, Bael, Ber, Custard apple
Frozen Puree	Bael, Karonda, Ker, Phalsa Tamarind, Custard apple
Canning	Ber, Aonla, Jamun, Ker

The underutilized fruits are hidden treasure of vital nutrients and have several pharmaceutical importance, some of them are used as a table fruit but majority of them have unique processing qualities, the only need is to tap this potential and convert them health into boosting products for common mass to make a viable option for cultivation by the farmers.

□□□